T0136874

Studies in Systems, Decision and Control

Volume 276

Series Editor

Janusz Kacprzyk, Systems Research Institute, Polish Academy of Sciences, Warsaw, Poland

The series "Studies in Systems, Decision and Control" (SSDC) covers both new developments and advances, as well as the state of the art, in the various areas of broadly perceived systems, decision making and control–quickly, up to date and with a high quality. The intent is to cover the theory, applications, and perspectives on the state of the art and future developments relevant to systems, decision making, control, complex processes and related areas, as embedded in the fields of engineering, computer science, physics, economics, social and life sciences, as well as the paradigms and methodologies behind them. The series contains monographs, textbooks, lecture notes and edited volumes in systems, decision making and control spanning the areas of Cyber-Physical Systems, Autonomous Systems, Sensor Networks, Control Systems, Energy Systems, Automotive Systems, Biological Systems, Vehicular Networking and Connected Vehicles, Aerospace Systems, Automation, Manufacturing, Smart Grids, Nonlinear Systems, Power Systems, Robotics, Social Systems, Economic Systems and other. Of particular value to both the contributors and the readership are the short publication timeframe and the world-wide distribution and exposure which enable both a wide and rapid dissemination of research output.

** Indexing: The books of this series are submitted to ISI, SCOPUS, DBLP, Ulrichs, MathSciNet, Current Mathematical Publications, Mathematical Reviews, Zentralblatt Math: MetaPress and Springerlink.

More information about this series at http://www.springer.com/series/13304

Martine Ceberio · Vladik Kreinovich
Editors

Decision Making under Constraints

 Springer

Editors
Martine Ceberio
Department of Computer Science
University of Texas at El Paso
El Paso, TX, USA

Vladik Kreinovich
Department of Computer Science
University of Texas at El Paso
El Paso, TX, USA

ISSN 2198-4182   ISSN 2198-4190   (electronic)
Studies in Systems, Decision and Control
ISBN 978-3-030-40816-9   ISBN 978-3-030-40814-5   (eBook)
https://doi.org/10.1007/978-3-030-40814-5

This Springer imprint is published by the registered company Springer Nature Switzerland AG
The registered company address is: Gewerbestrasse 11, 6330 Cham, Switzerland

Preface

In many application areas, it is necessary to make effective decisions under constraints. Several area-specific techniques are known for such decision problems; however, because these techniques are area-specific, it is not easy to apply each technique to other applications areas.

Cross-fertilization between different application areas is one of the main objectives of the annual International Workshops on Constraint Programming and Decision Making. Papers from the previous workshops appeared in [9] and in [10].

The 2016–2018 CoProd workshops, held in Europe (Upsala, Sweden), in America (El Paso, Texas), and in Asia (Tokyo, Japan), have also attracted researchers and practitioners from all over the world.

This volume presents extended versions of selected papers from these workshops—as well as several papers from the 2019 workshop which was co-located with the 2019 World Congress of the International Fuzzy Systems Association IFSA'2019 (Lafayette, Louisiana, USA).

These papers deal with all stages of decision making under constraints:

- formulating the problem of (in general, multi-criteria) decision making in precise terms [5, 6, 14, 27];
- determining when the corresponding decision problem is algorithmically solvable [8, 14, 16, 20];
- finding the corresponding algorithms, and making these algorithms as efficient as possible [2, 3, 5, 6, 7, 13, 15, 18, 20, 21, 24]; and
- taking into account interval uncertainty [8, 15, 16, 17, 18, 22, 23, 26], probabilistic uncertainty [1, 12, 15], and fuzzy uncertainty [1, 4, 11, 12, 15, 25] inherent in the corresponding decision making problems.

Several papers describe applications, in particular, applications:

- to biology [19],
- to engineering: to control of a smart thermostat [25], to control of Unmanned Aerial Vehicles [18], to power engineering [26], to structural mechanics [22], to vehicle protection against Improvised Explosion Devices [28], to waste water engineering [27],
- to finance [15],
- to software engineering [29].

We are greatly thankful to all the authors and referees, and to all the participants of the CoProd workshops. Our special thanks to Prof. Janusz Kacprzyk, the editor of this book series, for his support and help. Thanks to all of you!

El Paso, USA   Martine Ceberio
December 2019   mceberio@utep.edu

                                                                                        Vladik Kreinovich
                                                                                        vladik@utep.edu

References

1. Afravi M., Kreinovich V.: Fuzzy Systems Are Universal Approximators for Random Dependencies: A Simplified Proof (this volume)
2. Ayub, C., Ceberio, M., Kreinovich, V.: How quantum computing can help with (continuous) optimization (this volume)
3. Baral, C., Ceberio, M., Kreinovich, V.: How neural networks (NN) can (hopefully) learn faster by taking Into account known constraints (this volume)
4. Bergamaschi, F.B., Santiago, R.H.: Fuzzy primeness in quantales (this volume)
5. Bistarelli, S.: Collective defense and possible relaxations in weighted abstract argumentation problems (this volume)
6. Cao, Y., Ezawa, Y., Chen, G., Pan, H.: Modeling and specification of nondeterministic fuzzy discrete-event systems (this volume)
7. Ceberio, M., Kosheleva, O.,Kreinovich, V.: Italian folk multiplication algorithm Is indeed better: It is more parallelizable (this volume)
8. Ceberio, M., Kosheleva, O.,Kreinovich, V.: Reverse mathematics is computable for interval computations (this volume)
9. Ceberio, M., Kreinovich, V. (eds.): Constraint Programming and Decision Making, Springer, Berlin, Heidelberg (2014)
10. Ceberio, M., Kreinovich, V. (eds.): Constraint Programming and Decision Making: Theory and Applications, Springer, Cham, Switzerland, p. 208
11. Dvorak, A., Holcapek, M.: Generalized ordinal sum constructions of t-norms on bounded lattices (this volume)
12. Figueroa-Garcia, J.C.: A comparison of some t-norms and t-conorms over the steady state of a fuzzy markov chain (this volume)
13. Garcia Contreras, A.F., Ceberio, M., Kreinovich, V.: Plans are worthless but planning is everything: A theoretical explanation of eisenhower's observation (this volume)
14. Garcia Contreras, A.F., Ceberio, M., Kreinovich, V.: Why convex optimization is ubiquitous and why pessimism is widely spread (this volume)

Contents

Fuzzy Systems Are Universal Approximators for Random Dependencies: A Simplified Proof

Mahdokht Afravi and Vladik Kreinovich

Abstract In many real-life situations, we do not know the actual dependence $y = f(x_1, \ldots, x_n)$ between the physical quantities x_i and y, we only know expert rules describing this dependence. These rules are often described by using imprecise ("fuzzy") words from natural language. Fuzzy techniques have been invented with the purpose to translate these rules into a precise dependence $y = \widetilde{f}(x_1, \ldots, x_n)$. For deterministic dependencies $y = f(x_1, \ldots, x_n)$, there are universal approximation results according to which for each continuous function on a bounded domain and for every $\varepsilon > 0$, there exist fuzzy rules for which the resulting approximate dependence $\widetilde{f}(x_1, \ldots, x_n)$ is ε-close to the original function $f(x_1, \ldots, x_n)$. In practice, many dependencies are *random*, in the sense that for each combination of the values $x_1, \ldots, x_n$, we may get different values y with different probabilities. It has been proven that fuzzy systems are universal approximators for such random dependencies as well. However, the existing proofs are very complicated and not intuitive. In this paper, we provide a simplified proof of this universal approximation property.

1 Formulation of the Problem

It is important to determine dependencies. One of the main objectives of science is to find the state of the world and to predict the future state of the world—both in situations when we do not interfere and when we perform a certain action. The state of the world is usually characterized by the values of appropriate physical quantities.

For example:

- we would like to know the distance y to a distant star,
- we would like to predict tomorrow's temperature y at a given location, etc.

M. Afravi · V. Kreinovich (✉)
University of Texas at El Paso, El Paso, TX 79968, USA
e-mail: vladik@utep.edu

M. Afravi
e-mail: mafravi@miners.utep.edu

© Springer Nature Switzerland AG 2020
M. Ceberio and V. Kreinovich (eds.), *Decision Making under Constraints*,
Studies in Systems, Decision and Control 276,
https://doi.org/10.1007/978-3-030-40814-5_1

In some cases, we can directly measure the current value of the quantity y of interest. However, in many practical cases, such a direct measurement is not possible—e.g.:

- while it is possible to measure a distance to a nearby town by just driving there,
- it is not yet possible to directly travel to a faraway star.

And it is definitely not possible to measure tomorrow's temperature y today.

In such situations, since we cannot directly measure the value of the desired quantity y, a natural idea is:

- to measure related easier-to-measure quantities $x_1, \ldots, x_n$, and then
- to use the known dependence $y = f(x_1, \ldots, x_n)$ between these quantities to estimate y.

For example, to predict tomorrow's temperature at a given location, we can:

- measure today's values of temperature, wind velocity, humidity, etc. in nearby locations, and then
- use the known equations of atmospheric physics to predict tomorrow's temperature $\sim y$.

In some cases we know the exact form of the dependence $y = f(x_1, \ldots, x_n)$, but in many other practical situations, we do not have this information. Instead, we have to rely on experts who often formulate their rules in terms of imprecise ("fuzzy") words from natural language.

Imprecise ("fuzzy") rules and how they can be transformed into formulas. What kind of imprecise rules can we have? In some cases, the experts formulating the rule are imprecise both about x_i and about y. In such situations, we may have rules like this: "if today's temperature is very low and the Northern wind is strong, the temperature will remain very low tomorrow." In this case, x_1 is temperature today, x_2 is the speed of the Northern wind, y is tomorrow's temperature, and the properties "very low" and "strong" are imprecise.

In general, we have rules of the type

$$\text{"if } x_1 \text{ is } A_{k1}, \ldots, \text{ and } x_n \text{ is } A_{kn}, \text{ then } y \text{ is } A_k\text{"},$$

where A_{ki} and A_k are imprecise properties.

It is worth mentioning that in some cases, the information about x_i is imprecise, but the conclusion about y is described by a precise expression. For example, in non-linear mechanics, we can say that when the stress x_1 is small, the strain y is determined by a linear formula $y = k \cdot x_1$, with known k, but when the stress is high, we need to use a nonlinear expression $y = k \cdot x_1 - a \cdot x_1^2$ with known k and a. Here, both expressions are exactly known, but the condition when to apply one or another is described in terms of imprecise words like "small".

To transform such expert rules into a precise expression, Zadeh invented fuzzy logic; see, e.g., [1, 4, 5]. In fuzzy logic, to describe each imprecise property P, we ask the expert to assign, to each possible value x of the corresponding quantity, a degree $\mu_P(x)$ to which the value x satisfies this property—e.g., to what extent the value x is small. We can do this, e.g., by asking the expert to mark, on a scale from 0 to 10 to what extent the given value x is small. If the expert marks 7, we take $\mu_P(x) = 7/10$. The function $\mu_P(x)$ that assigns this degree is known as the *membership function* corresponding to the property P.

For given inputs $x_1, \ldots, x_n$, a value y is possible if it fits within one of the rules, i.e., if:

- either the first rule is satisfied, i.e., x_1 is A_{11}, ..., x_n is A_{1n}, and y is A_1,
- or the second rule is satisfied, i.e., x_1 is A_{21}, ..., x_n is A_{2n}, and y is A_2, etc.

Since we assumed that we know the membership functions $\mu_{ki}(x_i)$ and $\mu_k(y)$ corresponding to the properties A_{ki} and A_k, we can thus find the degrees $\mu_{ki}(x_i)$ and $\mu_k(y)$ to which each corresponding property is satisfied.

To estimate the degree to which y is possible, we must be able to deal with propositional connectives "or" and "and", i.e., to come up with a way to estimate our degrees of confidence in statements $A \vee B$ and $A \& B$ based on the known degrees of confidence a and b of the elementary statements A and B. In fuzzy logic, such estimation algorithms are known as *t-conorms* ("or"-operations) and *t-norms* ("and"-operations). We will denote them by $f_\vee(a, b)$ and $f_\&(a, b)$. In these terms, the degree $\mu(y)$ to which each value y is possible can be estimated as $\mu(y) = f_\vee(r_1, r_2, \ldots)$, where

$$r_k \overset{\text{def}}{=} f_\&(\mu_{k1}(x_1), \ldots, \mu_{kn}(x_n), \mu_k(y)).$$

We can then transform these degrees into a numerical estimate $\overline{y}$. This can be done, e.g., by minimizing the weighted mean square difference $\int \mu(y) \cdot (y - \overline{y})^2 \, dy$, which results in

$$\overline{y} = \frac{\int y \cdot \mu(y) \, dy}{\int \mu(y) \, dy}.$$

Universal approximation result for deterministic dependencies. For deterministic dependencies $y = f(x_1, \ldots, x_n)$, there are universal approximation results according to which for each continuous function on a bounded domain and for every $\varepsilon > 0$, there exist fuzzy rules for which the resulting approximate dependence $\widetilde{f}(x_1, \ldots, x_n)$ is ε-close to the original function $f(x_1, \ldots, x_n)$ for all the values x_i from the given domain.

In practice, we can often only make probabilistic predictions. In practice, many dependencies are *random*, in the sense that for each combination of the values $x_1, \ldots, x_n$, we may get different values y with different probabilities.

Fuzzy systems are universal approximators for random dependencies as well. It has been proven that fuzzy systems and universal approximators for random dependencies as well; see, e.g., [2, 3].

Remaining problem: can we simplify these proofs. The proofs presented in [2, 3] are very complicated and not intuitive. It is therefore desirable to simplify these proofs.

What we do in this paper. In this paper, we provide a simplified proof of the universal approximation property for random dependencies.

2   Towards a Simplified Proof

Main idea: how do we simulate random dependencies? To simulate a deterministic dependence $y = f(x_1, \ldots, x_n)$, we design an algorithm that, given the values $x_1, \ldots, x_n$, computes the corresponding value y.

To simulate a random dependence, a computer must also use the results of some *random number generators* that generate numbers distributed according to some probability distribution. Such generators are usually based on the basic random number generator—which is either supported by the corresponding programming language or even on a hardware level—that generates numbers uniformly distributed on the interval $[0, 1]$.

From this viewpoint, the result of simulating a random dependency has the form

$$y = F(x_1, \ldots, x_n, \omega_1, \ldots, \omega_m),$$

where F is the corresponding algorithm, x_i are inputs, and the values ω_j comes from the basic random number generator.

In these terns, what does it mean to approximate? In the above terms, to approximate means to find a function $\widetilde{F}$ for which, for all possible inputs x_i from the given bonded range, and for all possible values ω_j, the corresponding value

$$\widetilde{y} = \widetilde{F}(x_1, \ldots, x_n, \omega_1, \ldots, \omega_m)$$

are ε-close to the results of applying the algorithm F to the same values x_i and ω_j.

This leads to a simplified proof. The above idea leads to following simplified proof:

- due to the universal approximation theorem for deterministic dependencies, for every $\varepsilon > 0$, there exists a system of fuzzy rules for which the value of the corresponding function $\widetilde{F}$ is ε-close to the value of the original function F;
- thus, we get a fuzzy system of rules that provides the desired approximation to the original random dependency F.

Acknowledgements This work was supported in part by the National Science Foundation grants HRD-0734825 and HRD-1242122 (Cyber-ShARE Center of Excellence) and DUE-0926721, and by an award "UTEP and Prudential Actuarial Science Academy and Pipeline Initiative" from Prudential Foundation.

References

1. Klir, G., Yuan, B.: Fuzzy Sets and Fuzzy Logic. Prentice Hall, Upper Saddle River, New Jersey (1995)
2. Liu, P.: Mamdani fuzzy system: universal approximator to a class of random processes. IEEE Transactions on Fuzzy Systems **10**(6), 756–766 (2002)
3. Liu, P., Li, H.: Approximation of stochastic processes by T-S fuzzy systems. Fuzzy Sets and Systems **155**, 215–235 (2005)
4. Nguyen, H.T., Walker, E.A.: A First Course in Fuzzy Logic. Chapman and Hall/CRC, Boca Raton, Florida (2006)
5. Zadeh, L.A.: Fuzzy sets. Information and Control **8**, 338–353 (1965)

How Quantum Computing Can Help with (Continuous) Optimization

Christian Ayub, Martine Ceberio and Vladik Kreinovich

Abstract It is known that the use of quantum computing can reduce the time needed for a search in an unsorted array: from the original non-quantum time T to a much smaller quantum computation time $T_q \sim \sqrt{T}$. In this paper, we show that for a continuous optimization problem, with quantum computing, we can reach almost the same speed-up: namely, we can reduce the non-quantum time T to a much shorter quantum computation time $\sqrt{T} \cdot \ln(T)$.

1 Formulation of the Problem

Known advantages of quantum computing. It is known that quantum computing enables us to drastically speed up many computations; see, e.g., [3].

One example of such a problem is the problem of looking for a given element in an unsorted n-element array. With non-quantum computations, to be sure that we have found this element, we need to spend at least n computational steps. Indeed, if we use fewer than n steps, this would mean that we only look at less than n elements of the array—and thus, we may miss the element that we are looking for.

Grover's quantum-computing algorithm [1, 2] allows us to reduce the time needed to search for an element in an unsorted array of size n from the non-quantum lower bound n to a much faster time $c \cdot \sqrt{n}$.

In other words, we reduce the time needed for this task from the non-quantum time T to a much smaller quantum computation time T_q which is proportional to $\sqrt{T}$.

C. Ayub · M. Ceberio · V. Kreinovich (✉)
Department of Computer Science, University of Texas at El Paso,
El Paso, TX 79968, USA
e-mail: vladik@utep.edu

C. Ayub
e-mail: cayub@miners.utep.edu

M. Ceberio
e-mail: mceberio@utep.edu

© Springer Nature Switzerland AG 2020
M. Ceberio and V. Kreinovich (eds.), *Decision Making under Constraints*,
Studies in Systems, Decision and Control 276,
https://doi.org/10.1007/978-3-030-40814-5_2

Need to consider optimization problems. While search problems are ubiquitous, in many applications, we also need to solve continuous optimization problems. In such problems, we want to find an object or a strategy for which the given objective function attains its largest possible (or smallest possible) value.

An object is usually characterized by its parameters $x_1, \ldots, x_n$, for each of which we usually know the bounds $\underline{x}_i$ and $\overline{x}_i$, so that $\underline{x}_i \leq x_i \leq \overline{x}_i$. Let $f(x_1, \ldots, x_n)$ denote the value of the objective function corresponding to the parameters $x_1, \ldots, x_n$.

Comment. In most practical situations, the objective function is several (at least two) times continuously differentiable (smooth).

General optimization problem: idealized case. The idealized optimization problem means finding the values $x_1^{\text{opt}}, \ldots, x_n^{\text{opt}}$ for which the function $f(x_1, \ldots, x_n)$ attains its largest possible value on the box

$$B \stackrel{\text{def}}{=} [\underline{x}_1, \overline{x}_1] \times \cdots \times [\underline{x}_n, \overline{x}_n], \tag{1}$$

i.e., for which

$$f(x_1^{\text{opt}}, \ldots, x_n^{\text{opt}}) = \max_{(x_1, \ldots, x_n) \in B} f(x_1, \ldots, x_n). \tag{2}$$

General optimization problem: practical formulation. Of course, in practice, we can only attain values approximately. So, in practice, we are looking not for the absolute maximum, but rather for the values $x_1, \ldots, x_n$ which are maximal with given accuracy $\varepsilon > 0$.

In other words, we are interested in the following problem:

- we are given a function $f(x_1, \ldots, x_n)$ defined on a given box B, and we are given a real number ε;
- we want to find values $x_1^d, \ldots, x_n^d$ for which

$$f(x_1^d, \ldots, x_n^d) \geq \left(\max_{(x_1, \ldots, x_n) \in B} f(x_1, \ldots, x_n) \right) - \varepsilon.$$

What we do in this paper. In this paper, we show that the use of quantum computing can speed up the solution of this problem as well—and we show how exactly this problem can be sped up, from non-quantum computation time T to a much shorter quantum computation time $T_q \sim \sqrt{T} \cdot \ln(T)$.

2 How This Optimization Problem Is Solved Now

We consider only guaranteed global optimization algorithms. Of course, there are many semi-heuristic ways to solve the optimization problem. For example, we can start at some point $x = (x_1, \ldots, x_n)$ and use gradient techniques to reach a *local* maximum. However, these methods only lead to a local maximum. If we want to make sure that we reached the actual (*global*) maximum, we cannot skip some subdomains of the box B, we have to analyze all of them. How can we do it?

Non-quantum lower bound. Similarly to the search problem, we can find a natural lower bound on the time complexity of a non-quantum algorithm for global optimization.

Indeed, let us select some size $\delta > 0$ (that will be determined later), and let us divide each interval $[\underline{x}_i, \overline{x}_i]$ into $N_i \overset{\text{def}}{=} \dfrac{\overline{x}_i - \underline{x}_i}{\delta}$ subintervals of width δ.

This divides the whole box B into into

$$N = N_1 \cdot \ldots \cdot N_n = \prod_{i=1}^{n} \frac{\overline{x}_i - \underline{x}_i}{\delta} = \frac{V}{\delta^n} \tag{3}$$

subboxes, where

$$V \overset{\text{def}}{=} (\overline{x}_1 - \underline{x}_1) \cdot \ldots \cdot (\overline{x}_n - \underline{x}_n) \tag{4}$$

is the volume of the original box B.

We can have functions which are 0 everywhere except for one subbox at which this function grows to some value slightly larger than $\varepsilon > 0$—e.g., equal to $1.1 \cdot \varepsilon$. On this subbox, the function is approximately quadratic. If we have a bound S on the second derivative, we conclude that this function—which starts with 0 at a neighboring subbox—cannot grow faster than $S \cdot x^2$ on this subbox. Thus, to reach a value larger than ε, we need to select δ for which $S \cdot (\delta/2)^2 = 1.1 \cdot \varepsilon$, i.e., the value $\delta \sim \varepsilon^{1/2}$. For this value δ, we get $V/\delta^n \sim \varepsilon^{-(n/2)}$ subboxes.

If we do not explore some values of the optimized function at each of the subboxes, we may miss the subbox that contains the largest value—and thus, we will not be able to localize the point at which the function attains its maximum. Thus, to locate the global maximum, we need at least as many computation steps as there are subboxes—i.e., we need at least time $\sim \varepsilon^{-(n/2)}$.

This lower bound is reachable. Let us show that, similarly to search in unsorted array, this lower bound is reachable: there exists an algorithm that always locates the global maximum in time $\sim \varepsilon^{-(n/2)}$.

First stage of this algorithm: estimation on each subbox. Let us divide the box B into subboxes of linear size δ.

For each such subbox b, since each of its sides has size $\leq \delta$, each component x_i of each point $x = (x_1, \ldots, x_n) \in b$ differs from the corresponding component of the subbox's midpoint $x^m \overset{\text{def}}{=} (x_1^m, \ldots, x_n^m)$ by no more than $\delta/2$: $|\Delta x_i| \leq \delta/2$, where we denoted $\Delta x_i \overset{\text{def}}{=} x_i - x_i^m$. Thus, by using known formulas from calculus, we can conclude that for each point $x = (x_1, \ldots, x_n) \in b$, we have

$$f(x_1, \ldots, x_n) = f(x_1^m + \Delta x_1, \ldots, x_n^m + \Delta x_n) =$$

$$f(x_1^m, \ldots, x_n^m) + \sum_{i=1}^{n} c_i \cdot \Delta x_i + \sum_{i=1}^{n} \sum_{j=1}^{n} c_{ij} \cdot \Delta x_i \cdot \Delta x_j, \qquad (5)$$

where we denoted $c_i \overset{\text{def}}{=} \dfrac{\partial f}{\partial x_i}(x_1^m, \ldots, x_n^m)$ and $c_{ij} \overset{\text{def}}{=} \dfrac{\partial^2 f}{\partial x_i \partial x_j}(\xi_1, \ldots, \xi_n)$, for some values $(\xi_1, \ldots, \xi_n) \in b$.

We assumed that the function f is twice continuously differentiable. So, all its second derivatives are continuous, and thus, there exists a general bound S on all the values of all second derivatives: $|c_{ij}| \leq S$. Because of these bounds, the quadratic terms in the formula (5) are bounded by $n^2 \cdot S \cdot (\delta/2)^2 = O(\delta^2)$.

These estimations lead to a global piece-wise linear approximate function. By considering only linear terms on each subbox, we get an approximate piece-wise linear function $f_\approx(x_1, \ldots, x_n)$ which, on each subbox b, has a linear form

$$f_\approx(x_1, \ldots, x_n) = f(x_1^m, \ldots, x_n^m) + \sum_{i=1}^{n} c_i \cdot \Delta x_i. \qquad (6)$$

For each $x = (x_1, \ldots, x_n)$, we have

$$|f(x_1, \ldots, x_n) - f_\approx(x_1, \ldots, x_n)| \leq n^2 \cdot S \cdot (\delta/2)^2. \qquad (7)$$

Optimizing the approximate function. Let us find the point at which the approximate piece-wise linear function $f_\approx(x_1, \ldots, x_n)$ attains its maximum.

For each subbox

$$[x_1^m - \delta/2, x_1^m + \delta/2] \times \cdots \times [x_n^m - \delta/2, x_n^m + \delta/2], \qquad (8)$$

as one can easily see:

- the function $f_\approx(x_1, \ldots, x_n)$ is increasing with respect to each x_i when $c_i \geq 0$ and
- the function $f_\approx(x_1, \ldots, x_n)$ is decreasing with respect to x_i if $c_i \leq 0$.

Thus:

- when $c_i \geq 0$, the maximum of the function $f_\approx(x_1, \ldots, x_n)$ on this subbox is attained when $x_i = x_i^m + \delta/2$,
- when $c_i \leq 0$, the maximum of the function $f_\approx(x_1, \ldots, x_n)$ on this subbox is attained when $x_i = x_i^m - \delta/2$.

We can combine both cases by saying that the maximum is attained when $x_i = x_i^m + \text{sign}(c_i) \cdot (\delta/2)$, where $\text{sign}(x)$ is the sign of x (i.e., 1 if $x \geq 0$ and -1 otherwise).

We can repeat this procedure for each subbox, find the corresponding largest value on each subbox, and then find the largest of these values. The point

$x^M = (x_1^M, \ldots, x_n^M)$ at which this largest value is attained is thus the point at which the piece-wise linear function $f_\approx(x_1, \ldots, x_n)$ attains its maximum.

Proof of correctness. Let us show that the point $x^M = (x_1^M, \ldots, x_n^M)$ is indeed a solution to the given optimization problem.

Indeed, let $x^{\text{opt}} = (x_1^{\text{opt}}, \ldots, x_n^{\text{opt}})$ be a point where the original function $f(x_1, \ldots, x_n)$ attains its maximum. Since the approximate function attains its maximum at the point x^M, its value $f_\approx(x^M)$ at this point is larger than or equal to any other value. In particular, we have

$$f_\approx(x_1^M, \ldots, x_n^M) \geq f_\approx(x_1^{\text{opt}}, \ldots, x_n^{\text{opt}}). \tag{9}$$

Since the functions $f_\approx(x_1, \ldots, x_n)$ and $f(x_1, \ldots, x_n)$ are η-close, where we denoted

$$\eta \stackrel{\text{def}}{=} n^2 \cdot S \cdot (\delta/2)^2, \tag{10}$$

we conclude that

$$f(x_1^M, \ldots, x_n^M) \geq f_\approx(x_1^M, \ldots, x_n^M) - \eta \tag{11}$$

and

$$f_\approx(x_1^{\text{opt}}, \ldots, x_n^{\text{opt}}) \geq f(x_1^{\text{opt}}, \ldots, x_n^{\text{opt}}) - \eta = M - \eta. \tag{12}$$

From (9), (11), and (12), we conclude that

$$f(x_1^M, \ldots, x_n^M) \geq f_\approx(x_1^M, \ldots, x_n^M) - \eta \geq f_\approx(x_1^{\text{opt}}, \ldots, x_n^{\text{opt}}) - \eta \geq \\ (M - \eta) - \eta = M - 2\eta, \tag{13}$$

i.e., that

$$f(x_1^M, \ldots, x_n^M) \geq M - 2\eta. \tag{14}$$

Thus, for $\eta = \varepsilon/2$, we indeed get the solution to the original problem.

Selecting the appropriate value of the parameter δ. So, to solve the original problem with a given ε, we need to select the value δ for which

$$2n^2 \cdot S \cdot (\delta/2)^2 = \varepsilon, \tag{15}$$

i.e., the value $\delta = c \cdot \varepsilon^{1/2}$, for an appropriate constant c.

How much computation time do we need. In this algorithm, we divide the whole box B of volume V into V/δ^n subboxes of linear size δ. Since $\delta \sim \varepsilon^{1/2}$, the overall number of subboxes is proportional to $\varepsilon^{-n/2}$. On each subbox, the number of computational steps does not depend on ε, so the overall computation time is proportional to the number of boxes, i.e.,

$$T = \text{const} \cdot \varepsilon^{-n/2}. \tag{16}$$

3 How Quantum Computing Can Help

Preliminary step: bounding the approximate function. Similarly to how we bounded the function on a subbox, we can use the bound S on the second derivative and find the upper bound on the function $f(x_1, \ldots, x_n)$ over the whole box B, i.e., we can find the value $\overline{F}$ for which $f(x_1, \ldots, x_n) \leq \overline{F}$ for all $x = (x_1, \ldots, x_n) \in B$.

This value $\overline{F}$ also serves as the upper bound for the desired maximum M of the function $f(x_1, \ldots, x_n)$. As the lower bound for the maximum, we can take, e.g., the value $\underline{F}$ of the function $f(x_1, \ldots, x_n)$ at the midpoint of the box B. Thus, we know that the maximum M lies in the interval $[\underline{F}, \overline{F}]$.

By selecting an appropriate value $\delta \sim \varepsilon^{1/2}$, we can get an approximate function $f_{\approx}(x)$ which is $(\varepsilon/4)$-close to the original function $f(x)$. Because of this closeness, the maximum $M_{\approx}$ of the approximate function is, as we have shown in the previous section, $(\varepsilon/2)$-close to the maximum M and is, thus, located in the interval $[\underline{A_0}, \overline{A_0}]$, where $\underline{A_0} \stackrel{\text{def}}{=} \underline{F} - \varepsilon/2$ and $\overline{A_0} \stackrel{\text{def}}{=} \overline{F} + \varepsilon/2$.

Auxiliary quantum algorithm. For each rational value A, we can use Grover's algorithm to find, in time $\sim \sqrt{N}$, one of N subboxes at which the maximum of $f_{\approx}(x)$ on this subbox is larger than or equal to A (or that there is no such subbox).

How we will use the auxiliary quantum algorithm: main idea. Let us assume that we know an interval $[\underline{A}, \overline{A}]$ that contains the maximum $M_{\approx}$ of the approximate function. Let us apply the above auxiliary quantum algorithm for $A = (\underline{A} + \overline{A})/2$.

If, applying the auxiliary quantum algorithm, we find our that there is a subbox b for which $f_{\approx}(x_0) \geq A$ for some $x_0 \in b$, then we will be able to conclude that

$$M_{\approx} = \max_{x \in B} f_{\approx}(x) \geq f(x_0) \geq A \tag{17}$$

and thus, that $M_{\approx} \geq A$ and that the actual maximum $M_{\approx}$ of the function $f_{\approx}$ is located somewhere in the interval $[A, \overline{A}]$.

On the other hand, if, after applying the auxiliary quantum algorithm, we find out that no such subbox exists, this means that $f_{\approx}(x) \leq A$ for all x. Thus, the maximum $M_{\approx}$ of the auxiliary function $f_{\approx}(x)$ is also smaller than or equal to A. Hence, in this case, the actual maximum $M_{\approx}$ of the function $f_{\approx}$ is somewhere in the interval $[\underline{A}, A]$.

In both cases, we get an interval of half-size that contains the value $M_{\approx}$.

Main algorithm: first part. Now, we can run the following bisection algorithm.

We start with the interval $[\underline{A}, \overline{A}] = [\underline{A_0}, \overline{A_0}]$ that contains the actual value $M_{\approx}$.

At each iteration, we apply the above idea with $A = (\underline{A} + \overline{A})/2$, and as a result, we come up with a half-size interval containing $M_{\approx}$.

In k steps, we decrease the width of the interval 2^k times, to $2^{-k} \cdot (\overline{A} - \underline{A})$. In particular, in $k \approx \ln(\varepsilon)$ steps, we can get an interval $[\underline{a}, \overline{a}]$ containing $M_{\approx}$ whose width is $\leq \varepsilon/4$: $\underline{a} \leq M_{\approx} \leq \overline{a}$.

Main algorithm: second part. Since the value $\underline{a}$ is smaller than or equal to the maximum $M_\approx$ of the approximate function $f_\approx(x_1, \ldots, x_n)$, one of the values of this approximate function is indeed greater than or equal to $\underline{a}$.

The above-described auxiliary quantum algorithm will then find, in time $\sim \sqrt{N}$, such a point $x^q = (x_1^q, \ldots, x_n^q)$ for which $f_\approx(x_1^q, \ldots, x_n^q) \geq \underline{a}$.

Proof of correctness. Let us prove that the resulting point $x^q = (x_1^q, \ldots, x_n^q)$ indeed solves the original optimization problem.

Indeed, by the very construction of this point, the value $f_\approx(x^q)$ is greater than or equal to $\underline{a}$. Since the value $f_\approx(x^q)$ cannot exceed the maximum value $M_\approx$ of the approximate function, and this maximum value is $\leq \overline{a}$, we conclude that $f_\approx(x^q) \leq \overline{a}$. Thus, both $f_\approx(x^q)$ and $M_\approx$ belong to the same interval $[\underline{a}, \overline{a}]$ of width $\leq \varepsilon/4$. Thus, the value $f_\approx(x^q)$ is $(\varepsilon/4)$-close to the maximum $M_\approx$. In particular, this implies that

$$f_\approx(x_1^q, \ldots, x_n^q) \geq M_\approx - \varepsilon/4. \tag{18}$$

Since the functions $f_\approx(x_1, \ldots, x_n)$ and $f(x_1, \ldots, x_n)$ are $(\varepsilon/4)$-close, we can conclude that the maximum values $M_\approx$ and M of these two functions are also $(\varepsilon/4)$-close. In particular, this implies that

$$M_\approx \geq M - \varepsilon/4. \tag{19}$$

From (18) and (19), can conclude that

$$f(x_1^q, \ldots, x_n^q) \geq M - \varepsilon/2. \tag{20}$$

Since the functions are $(\varepsilon/4)$-close, we conclude that

$$f(x_1^q, \ldots, x_n^q) \geq f_\approx(x_1^q, \ldots, x_n^q) - \varepsilon/4 \tag{21}$$

and thus, that

$$f(x_1^q, \ldots, x_n^q) \geq f_\approx(x_1^q, \ldots, x_n^q) - \varepsilon/4 \geq (M - \varepsilon/2) - \varepsilon/4 > M - \varepsilon. \tag{22}$$

Thus, we indeed get the desired solution to the optimization problem.

What is the computational complexity of this quantum algorithm. How many computational steps do we need to implement this algorithm?

We need $\sim \ln(\varepsilon)$ iterations each of which requires time

$$\sim \sqrt{N} \sim \sqrt{\varepsilon^{-(n/2)}} = \varepsilon^{-(n/4)}. \tag{23}$$

Thus, the overall computation time T_q of this quantum algorithm is equal to

$$T_q \sim \varepsilon^{-(n/4)} \cdot \ln(\varepsilon). \tag{24}$$

How faster is this quantum algorithm than the non-quantum optimization? We know that the computation time T of the non-quantum algorithm is $T \sim \varepsilon^{-(n/2)}$; thus, $\varepsilon^{-(n/4)} \sim \sqrt{T}$.

Here, $\varepsilon \sim T^{-(2/n)}$, and thus, $\ln(\varepsilon) \sim \ln(T)$. Thus, we conclude that

$$T_q \sim \sqrt{T} \cdot \ln(T). \tag{25}$$

The main result is thus proven.

Acknowledgements This work was supported in part by the US National Science Foundation grant HRD-1242122 (Cyber-ShARE Center of Excellence).

The authors are thankful for all the participants of the NMSU/UTEP Workshop on Mathematics, Computer Science, and Computational Science (Las Cruces, New Mexico, April 6, 2019) for valuable suggestions.

References

1. Grover, L.K.: A fast quantum mechanical algorithm for database search. In: Proceedings of the 28th ACM Symposium on Theory of Computing, pp. 212–219 (1996)
2. Grover, L.K.: Quantum mechanics helps in searching for a needle in a haystack. Phys. Rev. Lett. **79**(2), 325–328 (1997)
3. Nielsen, M., Chuang, I.: Quantum Computation and Quantum Information. Cambridge University Press, Cambridge (2000)

How Neural Networks (NN) Can (Hopefully) Learn Faster by Taking into Account Known Constraints

Chitta Baral, Martine Ceberio and Vladik Kreinovich

Abstract Neural networks are a very successful machine learning technique. At present, deep (multi-layer) neural networks are the most successful among the known machine learning techniques. However, they still have some limitations, One of their main limitations is that their learning process still too slow. The major reason why learning in neural networks is slow is that neural networks are currently unable to take prior knowledge into account. As a result, they simply ignore this knowledge and simulate learning "from scratch". In this paper, we show how neural networks can take prior knowledge into account and thus, hopefully, learn faster.

1 Formulation of the Problem

Need for machine learning. In many practical situations, we know that the quantities $y_1, \ldots, y_L$ depend on the quantities $x_1, \ldots, x_n$, but we do not know the exact formula for this dependence. To get this formula, we measure the values of all these quantities in different situations $m = 1, \ldots, M$, and then use the corresponding measurement results $x_i^{(m)}$ and $y_\ell^{(m)}$ to find the corresponding dependence. Algorithms that "learn" the dependence from the measurement results are known as *machine learning* algorithms.

Neural networks (NN): main idea and successes. One of the most widely used machine learning techniques is the technique of *neural networks* (NN)—which is

C. Baral
Department of Computer Science, Arizona State University,
Tempe, AZ 85287-5406, USA
e-mail: chitta@asu.edu

M. Ceberio · V. Kreinovich (✉)
Department of Computer Science, University of Texas at El Paso,
El Paso, TX 79968, USA
e-mail: vladik@utep.edu

M. Ceberio
e-mail: mceberio@utep.edu

© Springer Nature Switzerland AG 2020
M. Ceberio and V. Kreinovich (eds.), *Decision Making under Constraints*,
Studies in Systems, Decision and Control 276,
https://doi.org/10.1007/978-3-030-40814-5_3

based on a (simplified) simulation of how actual neurons works in the human brain (a brief technical description of this technique is given in Sect. 2). This technique has many useful applications; see, e.g., [1, 2].

At present (2020) multi-layer ("deep") neural networks are, empirically, the most efficient of the known machine learning techniques.

Neural networks: limitations. One of the main limitations of neural networks is that their learning is very slow: they need many thousand iterations just to learn a simple dependence.

This slowness is easy to explain: the current neural networks always start "from scratch", from zero knowledge. In terms of simulating human brain, they do not simulate how we learn the corresponding dependence—they simulate how a newborn child will eventually learn to recognize this dependence. Of course, this inability to take any prior knowledge into account drastically slows down the learning process.

What is prior knowledge. Prior knowledge means that we know some relations ("constraints") between the desired values $y_1, \ldots, y_L$ and the observed values $x_1, \ldots, x_n$, i.e., we know several relations of the type

$$f_c(x_1, \ldots, x_n, y_1, \ldots, y_L) = 0, \quad 1 \le c \le C.$$

Prior knowledge helps humans learn faster. Prior knowledge helps us learn. Yes, it takes some time to learn this prior knowledge, but this has been done *before* we have samples of x_i and y_ℓ. As a result, the time from gathering the samples to generating the desired dependence decreases.

This is not simply a matter of accounting: the same prior knowledge can be used (and usually is used) in learning several different dependencies. For example, our knowledge of sines, logarithms, of calculus helps in finding the proper dependence in many different situations. So, when we learn the prior knowledge first, we decrease the overall time needed to learn all these dependencies.

How to speed up artificial neural networks: a natural idea. In view of the above explanation, a natural idea is to enable neural networks to take prior knowledge into account. In other words, instead of learning all the data "from scratch", we should first learn the constraints. Then, when it is time to use the data, we should be able to use these constraints to "guide" the neural network in the right direction.

What we do in this paper. In this paper, we show how to implement this idea and thus, how to (hopefully) achieve the corresponding speed-up.

To describe this idea, we first, in Sect. 2, recall how the usual NN works. Then, in Sect. 3, we show how we can perform a preliminary training of a NN, so that it can learn to satisfy the given constraints. Finally, in Sect. 4, we show how to train the resulting pre-trained NN in such a way that the constraints remain satisfied.

2 Neural Networks: A Brief Reminder

Signals in a biological neural network. In a biological neural network, a signal is represented by a sequence of spikes. All these spikes are largely the same, what is different is how frequently the spikes come.

Several sensor cells generate such sequences: e.g., there are cells that translate the optical signal into spikes, there are cells that translate the acoustic signal into spikes. For all such cells, the more intense the original physical signal, the more spikes per unit time it generates. Thus, the frequency of the spikes can serve as a measure of the strength of the original signal.

From this viewpoint, at each point in a biological neural network, at each moment of time, the signal can be described by a single number: namely, by the frequency of the corresponding spikes.

What is a biological neuron: a brief description. A biological neuron has several inputs and one output. Usually, spikes from different inputs simply get together—probably after some filtering. Filtering means that we suppress a certain proportion of spikes. If we start with an input signal x_i, then, after such a filtering, we get a decreased signal $w_i \cdot x_i$. Once all the inputs signals are combined, we have the resulting signal $\sum_{i=1}^{n} w_i \cdot x_i$.

A biological neuron usually has some excitation level w_0, so that if the overall input signal is below w_0, there is practically no output. The intensity of the output signal thus depends on the difference $d \stackrel{\text{def}}{=} \sum_{i=1}^{n} w_i \cdot x_i - w_0$. Some neurons are linear, their output is proportional to this difference. Other neurons are non-linear, they output is equal to $s_0(d)$ for some non-linear function $s_0(z)$. Empirically, it was found that the corresponding non-linear transformation takes the form $s_0(z) = 1/(1 + \exp(-z))$.

Comment. It should be mentioned that this is a simplified description of a biological neuron: the actual neuron is a complex *dynamical* system, in the sense that its output at a given moment of time depends not only on the current inputs, but also on the previous input values.

Artificial neural networks and how they learn. If we need to predict the values of several outputs $y_1, \ldots, y_\ell, \ldots, y_L$, then for each output y_ℓ, we train a separate neural network.

In an artificial neural networks, input signals $x_1, \ldots, x_n$ first go to the neurons of the first layer, then the results go to neurons of the second layer, etc.

In the simplest (and most widely used) arrangement, the second layer has linear neurons. In this arrangement, the neurons from the first layer produce the signals $y_{\ell,k} = s_0 \left(\sum_{i=1}^{n} w_{\ell,ki} \cdot x_i - w_{\ell,k0} \right), 1 \leq k \leq K_\ell$, which are then combined into an output $y_\ell = \sum_{k=1}^{K} W_{\ell,k} \cdot y_k - W_{\ell,0}$. This is called *forward propagation*. (In this paper, we will only describe formulas for this arrangement, since formulas for the multi-layer neural networks can be obtained by using the same idea.)

How a NN learns: derivation of the formulas. Once we have an observation $(x_1^{(m)}, \ldots, x_n^{(m)}, y_\ell^{(m)})$, we first input the values $x_1^{(m)}, \ldots, x_n^{(m)}$ into the current NN; the network generates some output $y_{\ell,NN}$. In general, this output is different from the observed output $y_\ell^{(m)}$. We therefore want to modify the weights $W_{\ell,k}$ and $w_{\ell,ki}$ so as to minimize the squared difference $J \stackrel{\text{def}}{=} (\Delta y_\ell)^2$, where $\Delta y_\ell \stackrel{\text{def}}{=} y_{\ell,NN} - y_\ell^{(m)}$. This minimization is done by using gradient descent, where each of the unknown values is updated as $W_{\ell,k} \to W_{\ell,k} - \lambda \cdot \dfrac{\partial J}{\partial W_{\ell,k}}$ and $w_{\ell,ki} \to w_{\ell,ki} - \lambda \cdot \dfrac{\partial J}{\partial w_{\ell,ki}}$. The resulting algorithm for updating the weights is known as *backpropagation*. This algorithm is based on the following idea.

First, one can easily check that $\dfrac{\partial J}{\partial W_{\ell,0}} = -2\Delta y$, so $\Delta W_{\ell,0} = -\lambda \cdot \dfrac{\partial J}{\partial W_{\ell,0}} = \alpha \cdot \Delta y_\ell$, where $\alpha \stackrel{\text{def}}{=} 2\lambda$. Similarly, $\dfrac{\partial J}{\partial W_{\ell,k}} = 2\Delta y_\ell \cdot y_{\ell,k}$, so $\Delta W_{\ell,k} = -\lambda \cdot \dfrac{\partial J}{\partial W_{\ell,k}} = 2\lambda \cdot \Delta y_\ell \cdot y_{\ell,k}$, i.e., $\Delta W_{\ell,k} = -\Delta W_{\ell,0} \cdot y_{\ell,k}$.

The only dependence of y_ℓ on $w_{\ell,ki}$ is via the dependence of $y_{\ell,k}$ on $w_{\ell,ki}$. So, for $w_{\ell,k0}$, we can use the chain rule and get $\dfrac{\partial J}{\partial w_{\ell,k0}} = \dfrac{\partial J}{\partial y_{\ell,k}} \cdot \dfrac{\partial y_{\ell,k}}{\partial w_{\ell,k0}}$, hence:

$$\frac{\partial J}{\partial w_{\ell,k0}} = 2\Delta y_\ell \cdot W_{\ell,k} \cdot s_0'\left(\sum_{i=1}^{n} w_{\ell,ki} \cdot x_i - w_{\ell,k0}\right) \cdot (-1).$$

For $s_0(z) = 1/(1 + \exp(-z))$, we have $s_0'(z) = \exp(-z)/(1 + \exp(-z))^2$, i.e.,

$$s_0'(z) = \frac{\exp(-z)}{1 + \exp(-z)} \cdot \frac{1}{1 + \exp(-z)} = s_0(z) \cdot (1 - s_0(z)).$$

Thus, in the above formula, where $s_0(z) = y_{\ell,k}$, we get $s_0'(z) = y_{\ell,k} \cdot (1 - y_{\ell,k})$, $\dfrac{\partial J}{\partial w_{\ell,k0}} = -2\Delta y_\ell \cdot W_{\ell,k} \cdot y_{\ell,k} \cdot (1 - y_{\ell,k})$, and

$$\Delta w_{\ell,k0} = -\lambda \cdot \frac{\partial J}{\partial w_{\ell,k0}} = \lambda \cdot 2\Delta y_\ell \cdot W_{\ell,k} \cdot y_{\ell,k} \cdot (1 - y_{\ell,k}).$$

So, we have $\Delta w_{\ell,k0} = -\Delta W_{\ell,k} \cdot W_{\ell,k} \cdot (1 - y_{\ell,k})$. For $w_{\ell,ki}$, we have

$$\frac{\partial J}{\partial w_{\ell,ki}} = 2\Delta y_\ell \cdot W_{\ell,k} \cdot y_{\ell,k} \cdot (1 - y_{\ell,k}) \cdot x_i = -\frac{\partial J}{\partial w_{\ell,k0}} \cdot x_i,$$

hence $\Delta w_{\ell,ki} = -x_i \cdot \Delta w_{\ell,k0}$. Thus, we arrive at the following algorithm:

Resulting algorithm. We pick some value α, and cycle through observations $(x_1, \ldots, x_n)$ with the desired outputs y_ℓ. For each observation, we first apply the forward propagation to compute the network's prediction $y_{\ell, NN}$, then we compute $\Delta y_\ell = y_{\ell, NN} - y_\ell$, $\Delta W_{\ell,0} = \alpha \cdot \Delta y_\ell$, $\Delta W_{\ell,k} = -\Delta W_{\ell,0} \cdot y_{\ell,k}$, $\Delta w_{\ell,k0} = -\Delta W_{\ell,k} \cdot W_{\ell,k} \cdot (1 - y_{\ell,k})$, and $\Delta w_{\ell,ki} = -\Delta w_{\ell,k0} \cdot x_i$, and update each weight w to $w_{\text{new}} = w + \Delta w$. We repeat this procedure until the process converges.

3 How to Pre-Train a NN to Satisfies Given Constraints

To train the network, we can use any observations $(x_1^{(m)}, \ldots, x_n^{(m)}, y_1^{(m)}, \ldots, y_L^{(m)})$ that satisfy all the known constraints.

To satisfy the constraints $f_c(x_1, \ldots, x_n, y_1, \ldots, y_L) = 0$, $1 \leq c \leq C$, means to minimize the distance from the vector of values $(f_1, \ldots, f_C)$ to the ideal point $(0, \ldots, 0)$, i.e., equivalently, to minimize the sum $F \overset{\text{def}}{=} \sum_{c=1}^{C} (f_c(x_1, \ldots, x_n, y_1, \ldots, y_L))^2$. To minimize this sum, we can use a similar gradient descent idea. From the mathematical viewpoint, the only difference from the usual backpropagation is the first step: here,

$$\frac{\partial F}{\partial W_{\ell,0}} = 2 \cdot \sum_{c=1}^{C} f_c \cdot \frac{\partial f_c}{\partial y_\ell}, \quad \text{hence} \quad \Delta W_{\ell,0} = -\alpha \cdot \sum_{c=1}^{C} f_c \cdot \frac{\partial f_c}{\partial y_\ell}.$$

Once we have computed $\Delta W_{\ell,0}$, all the other changes $\Delta W_{\ell,k}$ and $\Delta w_{\ell,ki}$ are computed based on the same formulas as above.

The consequence of this algorithm modification is that instead of L independent neural networks used to train each of the L outputs y_ℓ, now we have L dependent ones. The dependence comes from the fact that, to start a new cycle for each ℓ, we need to know the values $y_1, \ldots, y_L$ corresponding to all the outputs.

4 How to Retain Constraints When Training Neural Networks on Real Data

Once the networks is pre-trained so that the constraints are all satisfied, we need to train it on the real data. In this real-data training, we need to make sure that not only all the given data points fit, but that also all C constraints remain satisfied. In other words, on each step, we need to make sure not only that Δy_ℓ is close to 0, but also that $f_c(x_1, \ldots, x_n, y_1, \ldots, y_L)$ is close to 0 for all ℓ. So, similar to the previous section, instead of minimizing $J = (\Delta y_\ell)^2$, we should minimize a combined objective function $G \overset{\text{def}}{=} J + N \cdot F$, where N is an appropriate constant, and $F = \sum_{c=1}^{C} f_c^2$.

Similarly to pre-training, the only difference from the usual backpropagation algorithm is that we compute the values $\Delta W_{\ell,0}$ differently:

$$\Delta W_{\ell,0} = \alpha \cdot \left(\Delta y_\ell - N \cdot \sum_{c=1}^{C} f_c \cdot \frac{\partial f_c}{\partial y_\ell} \right).$$

Acknowledgements This work was supported in part by NSF grants HRD-0734825, HRD-1242122, and DUE-0926721, and by an award from Prudential Foundation.

References

1. Bishop, C.M.: Pattern Recognition and Machine Learning. Springer, N.Y. (2006)
2. Hinton, G.E., Osindero, S., Teh, Y.-W.: A fast learning algorithm for deep belief nets. Neural Comput. **18**, 1527–1554 (2006)

Fuzzy Primeness in Quantales

Flaulles Boone Bergamaschi and Regivan H. N. Santiago

Abstract This paper is an investigation about primeness in quantales environment. It is proposed a new definition for prime ideal in noncommutative setting. As a consequence, fuzzy primeness can be defined in similar way to ring theory.

1 Introduction

In 2013, Yang and Xu [1] published a paper about roughness in quantales (see Definition 1). In this paper the authors defined a prime ideal (see Ring Theory) based on elements of a quantale. After that they built the rough prime ideal in quantales over this concept. In 2014, Luo and Wang [2] used the same definition of prime ideals of quantales to write an investigation called roughness and fuzziness where the first ideas on semi-prime, primary and strong primeness are presented. As it is known, ideals are the main object in the investigation of ring theory and provide important information about the rings because they are structural pieces. The same may occur in quantales. The definition of prime ideals proposed in [1, 2] is based on elements of a quantale and we ponder it is geared to commutative environment. When we move from commutative to the noncommutative setting, elementwise should be replaced by an approach based on ideals. Nevertheless, some authors defined the concept of primeness for commutative and noncommutative cases without realizing that this concept may not be suitable for noncommutative setting as it was well shown by Navarro et al. in [3]. We state that the concept of prime ideal of general quantales could be defined as it is done in ring theory, i.e. based on ideals. The concept of prime ideal provided for quantales by Lingyun Yang and Luoshan Xu is more suitable for commutative quantales. Therefore, this paper provides a new concept

F. B. Bergamaschi (✉)
Southwest Bahia State University, Vitória da Conquista, Brazil
e-mail: flaulles@yahoo.com.br

R. H. N. Santiago
Federal University of Rio Grande do Norte, Natal, Brazil
e-mail: regivan@dimap.ufrn.br

© Springer Nature Switzerland AG 2020
M. Ceberio and V. Kreinovich (eds.), *Decision Making under Constraints*,
Studies in Systems, Decision and Control 276,
https://doi.org/10.1007/978-3-030-40814-5_4

of prime ideal for a general (commutative and noncommutative) quantale which the elementwise prime ideal definition proposed by Lingyun Yang and Luoshan Xu is called completely prime ideal. Thus, this paper shows the difference between a commutative quantale e noncommutative quantale, and a definition more appropriate for primeness in quantales.

The first aim of this paper is to study the notion of primeness in the following perspective: we rename prime ideal defined in [1] to *completely prime ideal* and define a new concept of prime ideal for quantales. Then we translate an important result in ring theory for quantales environment (Theorem 2) to prove that these two concepts coincide in the commutative setting, but are no longer valid in the noncommutative setting (see Proposition 3). Also, based on the studies of Lawrence and Handelman [4], started in 1975, the notion of strong primeness is developed for general quantales.

The second aim is to propose the concept of fuzzy primeness and fuzzy strong primeness as well as fuzzy uniform strong primeness for quantales following the ideas of Bergamaschi and Santiago [5, 6].

Finally, we introduce initial ideas of t-systems and m-systems of quantales as an alternative to deal with primeness. As we will see later an ideal is prime iff its complement is an m-system.

This paper has the following structure: Sect. 2 provides the definition of prime and completely prime ideal of a general quantale. The main parts of this section are Theorem 2, Proposition 4 and Proposition 3; Sect. 3 provides the concept of strong and uniform strong primeness in quantales where the t- and m-systems are introduced; Sect. 4 introduces the concept of primeness and uniform strong primeness for fuzzy ideals in quantales. The compatibility with α-cuts is also proved; and Sect. 5 provides the final remarks.

2   Basic Definitions

The definition of prime ideal used in [1] and [2] will be called herein completely prime ideals. We drew attention to the Theorem 2 where prime ideals can be characterized in a certain way via elements. But, as we will see, this characterization is more appropriate for commutative environment. The Proposition 4 shows that in the commutative case, prime and completely prime concepts coincide, which are no longer valid in the noncommutative setting according to Proposition 3. Finally, the concept of quantale prime is proposed.

Definition 1 [7] A quantale is a complete lattice Q with an associative binary operation $\circ$ satisfying:

$$a \circ \left(\bigvee_{k \in K} b_k \right) = \bigvee_{k \in K} (a \circ b_k), \quad \left(\bigvee_{k \in K} a_k \right) \circ b = \bigvee_{k \in K} (a_k \circ b)$$

for all $a, b, a_k, b_k \in Q$ and $k \in K$.

A quantale Q is called commutative whenever $a \circ b = b \circ a$ for $a, b \in Q$. In this paper we denote the least and greatest elements of a quantale by $\bot$ and $\top$ respectively. If there exists an element e in Q such that $x \circ e = e \circ x = x$ for all x in Q the quantale is called a quantale with identity. In this paper we consider quantales with identity.

Definition 2 [2] Let Q be a quantale. A non-empty subset $I \subseteq Q$ is called a right ideal of Q if it satisfies the following conditions:

(i) $a, b \in I$ implies $a \vee b \in I$;
(ii) for all $a, b \in Q, a \in I$ and $b \leq a$ imply $b \in I$,
(iii) for all $x \in Q$ and $a \in I$, we have $a \circ x \in I$.

Similarly we may define left ideal replacing (iii) by: (iii$'$) for all $x \in Q$ and $a \in I$, we have $x \circ a \in I$. If I is both right and left ideal of Q, we call I a two-sided ideal or simply an ideal of Q.

Clearly by (ii) $\bot \in I$. Also, the set of all ideals of Q is closed under arbitrary intersections.

In Q we denote the subset $I \circ J = \{i \circ j \in Q : i \in I \text{ and } j \in J\}$ and $A \vee B = \{a \vee b : a \in A \text{ and } b \in B\}$. Since the operation $\circ$ is associative, we have $(A \circ B) \circ C = A \circ (B \circ C)$. Also, if A is an two-sided ideal, then $A \circ Q, Q \circ A, Q \circ A \circ Q \subseteq A$.

As usual, $\vee$ induces an order relation $\leq$ on Q by putting $x \leq y \Leftrightarrow x \vee y = y$. Moreover, $\leq$ is a congruence i.e. for every $x, y, u, v \in Q$ if $x \leq y$ and $u \leq v$, then $x \circ u \leq y \circ v$. To prove this, we first observe that if $w \leq z$ then, for any $s \in Q$, $s \circ w \leq s \circ z$ and $w \circ s \leq z \circ s$ because $z = w \vee z$ implies $s \circ z = s \circ (w \vee z) = (s \circ w) \vee (s \circ z)$ and $z \circ s = (w \vee z) \circ s = (w \circ s) \vee (z \circ s)$; now suppose $x \leq y$ and $u \leq v$, then $x \circ u \leq y \circ u$ and $y \circ u \leq y \circ v$. Hence, $x \circ u \leq y \circ v$ by transitivity.

In what follows we propose a more general definition of prime ideals which encompasses commutative and non-commutative quantales.

Definition 3 A prime ideal in a quantale Q is any proper ideal P ($P \subset Q$ and $P \neq Q$) such that, whenever I, J are ideals of Q with $I \circ J \subseteq P$, either $I \subseteq P$ or $J \subseteq P$.

Definition 4 A subset P of a quantale Q is called completely prime ideal if whenever x and y are two elements of Q such that their product $x \circ y \in I$, then $x \in I$ or $y \in I$.

As we will see the concept of prime and completely prime ideals are different and coincide whenever Q is commutative.

Proposition 1 *If P is completely prime, then P is prime.*

Proof Suppose that P is completely prime and $I \circ J \subseteq P$, but $J \nsubseteq P$, where I, J are ideals of Q. Thus, there exists $j \in J$ such that $j \notin P$. For all $i \in I$ we have $i \circ j \in I \circ J \subseteq P$, as P is completely prime and $j \notin P$, then $i \in P$. Therefore $I \subseteq P$.

The Proposition 3 will show that the converse of this Proposition is not true.

Definition 5 [2] Let Q be a quantale and $A \subseteq Q$. The least ideal containing A is called the ideal generated by A, and denoted as $\langle A \rangle$.

Clearly, $\langle \emptyset \rangle = \{\bot\}$. If $\emptyset \neq A \subseteq Q$, then we have the following result.

Proposition 2 [2] *Let A be a non-empty subset of a quantale Q. Then $\langle A \rangle = \{x \in Q : x \leq \bigvee_{i=1}^{n} a_i, \text{ for some positive integer } n \text{ and } a_1, \ldots, a_n \in A \cup (A \circ Q) \cup (Q \circ A) \cup (Q \circ A \circ Q)\}$.*

We may denote $\langle a \rangle = \langle \{a\} \rangle$ and $a \circ Q = \{a\} \circ Q$.

Lemma 1 $\langle a \rangle \circ Q \subseteq \langle a \rangle$ *for all $a \in Q$. If there exists an unit 1 in Q, then $\langle a \rangle \circ Q = \langle a \rangle$.*

Proof Let $x \circ q \in \langle a \rangle \circ Q$, where $x \in \langle a \rangle$ and $q \in Q$. Hence, $x \circ q \leq \bigvee_{i=1}^{n}(a_i \circ q)$, where $a_i \circ q \in a \circ Q \cup Q \circ a \cup Q \circ a \circ Q$. Thus, $x \circ q \in \langle a \rangle$. On the other hand if there exists unit 1 in Q, we write $z \in \langle a \rangle$ as $z = z \circ 1$. Thus, $z \in \langle a \rangle \circ Q$ and we have $\langle a \rangle \circ Q = \langle a \rangle$.

Theorem 2 *For an ideal P in Q the following statements are equivalent:*

(1) P is prime ideal;
(2) $\langle a \rangle \circ \langle b \rangle \subseteq P$ implies $a \in P$ or $b \in P$;
(3) $a \circ Q \circ b \subseteq P$ implies $a \in P$ or $b \in P$.

Proof For (1) $\Rightarrow$ (2) note that $\langle a \rangle$ and $\langle b \rangle$ are ideals of Q. As P is prime and $\langle a \rangle \circ \langle b \rangle \subseteq P$, then $\langle a \rangle \subseteq P$ or $\langle b \rangle \subseteq P$. Hence, $a \in P$ or $b \in P$. For (2) $\Rightarrow$ (1), assume that $I \circ J \subseteq P$, but $J \not\subseteq P$, where I, J are ideals of Q. Thus, there exists $j \in J$ such that $j \notin P$. Given $i \in I$ we have $\langle i \rangle \subseteq I$. Hence, $\langle i \rangle \circ \langle j \rangle \subseteq I \circ J \subseteq P$. By hypothesis $i \in P$ or $j \in P$. As $j \notin P$ then we have $i \in P$. Therefore, $I \subseteq P$.

For (3) $\Rightarrow$ (1), assume that $I \circ J \subseteq P$, but $J \not\subseteq P$, where I, J are ideals of Q. Thus, there exists $j \in J$ such that $j \notin P$. Given $i \in I$ we have $i \circ Q \circ j \subseteq I \circ J \subseteq P$. Hence, $i \in P$ or $j \in P$, as $j \notin P$ then we have $i \in P$. Thus, $I \subseteq P$.

For (1) $\Rightarrow$ (3), suppose $a \circ Q \circ b \subseteq P$, we first shall show that $\langle a \rangle \circ Q \circ \langle b \rangle \subseteq P$. For this, let $x \circ q \circ y \in \langle a \rangle \circ Q \circ \langle b \rangle$, where $x \in \langle a \rangle$, $y \in \langle b \rangle$ and $q \in Q$. Hence, by Proposition 2, $x \leq \bigvee_{i=1}^{n} a_i$ and $y \leq \bigvee_{j=1}^{m} b_j$, where $a_i \in (a \circ Q) \cup (Q \circ a) \cup (Q \circ a \circ Q)$ and $b_i \in (b \circ Q) \cup (Q \circ b) \cup (Q \circ b \circ Q)$. Hence $x \circ q \circ y \leq \left(\bigvee_{i=1}^{n} a_i \right) \circ q \circ \left(\bigvee_{j=1}^{m} b_j \right) = \left(\bigvee_{i=1}^{n}(a_i \circ q) \right) \circ \left(\bigvee_{j=1}^{m} b_j \right) = \bigvee_{i=1}^{n}(a_i \circ q \circ \bigvee_{j=1}^{m} b_j) = \bigvee_{i=1}^{n} \left(\bigvee_{j=1}^{m}(a_i \circ q \circ b_j) \right)$.

Observe that $a_i \in a \circ Q \cup Q \circ a \cup Q \circ a \circ Q$ and $b_j \in b \circ Q \cup Q \circ b \cup Q \circ b \circ Q$ it is no hard to see that $a_i \circ q \circ b_j \in a \circ Q \circ b \subseteq P$ for all i, j. As P is an ideal we have $x \circ q \circ y \in P$. Thus, $\langle a \rangle \circ Q \circ \langle b \rangle \subseteq P$. By the Lemma 1 $\langle a \rangle \circ Q = \langle a \rangle$. Then, $\langle a \rangle \circ Q \circ \langle b \rangle = \langle a \rangle \circ \langle b \rangle \subseteq P$. By the first proof ((1) $\Leftrightarrow$ (2)) we have $a \in P$ or $b \in P$.

Proposition 3 *There exists a noncommutative quantale where a prime ideal is not a completely prime ideal.*

Proof Consider G the set of all invertible 2×2 matrices under multiplication over the real interval $[0, 1]$ and the partial order $A \leq B \Leftrightarrow a_{ij} \leq b_{ij}$. According to Rosenthal ([7], p. 19, Example 16) any complete partially ordered group (written multiplicatively) is a quantale with $a \circ b = a \cdot b$. Thus, G is a noncommutative quantale.

Let $\langle 0 \rangle$ as an ideal generated by 0, clearly $\langle 0 \rangle = \{0\}$. We will show that the $\langle 0 \rangle$ (zero ideal) is prime, but not completely prime by using the Theorem 2 (3). Thus, suppose that $X = \begin{pmatrix} a & b \\ c & d \end{pmatrix}$ and $Y = \begin{pmatrix} e & f \\ g & h \end{pmatrix}$ are two matrices such that $X \circ G \circ Y \subseteq \langle 0 \rangle$.

Hence $X \circ T \circ Y = \begin{pmatrix} 0 & 0 \\ 0 & 0 \end{pmatrix}$ for all matrix $T \in G$. Then, in particular,

$$X \circ \begin{pmatrix} 1 & 0 \\ 0 & 0 \end{pmatrix} \circ Y = \begin{pmatrix} a & b \\ c & d \end{pmatrix} \begin{pmatrix} 1 & 0 \\ 0 & 0 \end{pmatrix} \begin{pmatrix} e & f \\ g & h \end{pmatrix} = \begin{pmatrix} ae & af \\ ce & cf \end{pmatrix} = 0 \Leftrightarrow a = c = 0 \text{ or}$$
$e = f = 0$,

$$X \begin{pmatrix} 0 & 1 \\ 0 & 0 \end{pmatrix} Y = \begin{pmatrix} a & b \\ c & d \end{pmatrix} \begin{pmatrix} 0 & 1 \\ 0 & 0 \end{pmatrix} \begin{pmatrix} e & f \\ g & h \end{pmatrix} = \begin{pmatrix} ag & ah \\ cg & ch \end{pmatrix} = 0 \Leftrightarrow a = c = 0 \text{ or}$$
$g = h = 0$,

$$X \circ \begin{pmatrix} 0 & 0 \\ 1 & 0 \end{pmatrix} \circ Y = \begin{pmatrix} a & b \\ c & d \end{pmatrix} \begin{pmatrix} 0 & 0 \\ 1 & 0 \end{pmatrix} \begin{pmatrix} e & f \\ g & h \end{pmatrix} = \begin{pmatrix} be & bf \\ de & df \end{pmatrix} = 0 \Leftrightarrow b = d = 0 \text{ or}$$
$e = f = 0$,

$$X \circ \begin{pmatrix} 0 & 0 \\ 0 & 1 \end{pmatrix} \circ Y = \begin{pmatrix} a & b \\ c & d \end{pmatrix} \begin{pmatrix} 0 & 0 \\ 0 & 1 \end{pmatrix} \begin{pmatrix} e & f \\ g & h \end{pmatrix} = \begin{pmatrix} bg & bh \\ dg & dh \end{pmatrix} = 0 \Leftrightarrow b = d = 0 \text{ or}$$
$g = h = 0$,

Hence, a solution must verify that $X = \begin{pmatrix} 0 & 0 \\ 0 & 0 \end{pmatrix}$ or $Y = \begin{pmatrix} 0 & 0 \\ 0 & 0 \end{pmatrix}$. Therefore $X \in \langle 0 \rangle$ or $Y \in \langle 0 \rangle$ and then $\langle 0 \rangle$ is prime. Nevertheless, $\langle 0 \rangle$ is not completely prime, since $\begin{pmatrix} 0 & 1 \\ 0 & 0 \end{pmatrix} \circ \begin{pmatrix} 0 & 1 \\ 0 & 0 \end{pmatrix} = \begin{pmatrix} 0 & 0 \\ 0 & 0 \end{pmatrix}$ although $\begin{pmatrix} 0 & 1 \\ 0 & 0 \end{pmatrix} \notin \langle 0 \rangle$.

Proposition 4 *In a commutative quantale an ideal is completely prime iff it is prime.*

Proof If P is a completely prime ideal of a quantale Q, then by the Proposition 1 P is prime. On the other hand, suppose P is a prime ideal and $a \circ b \in P$ for any $a, b \in Q$. Let $x \circ y \in \langle a \rangle \circ \langle b \rangle$, where $x \in \langle a \rangle$ and $y \in \langle b \rangle$. Thus, $x \circ y \leq \bigvee_{i=1}^{n} a_i \circ \bigvee_{j=1}^{m} b_j = \bigvee_{i=1}^{n} \left(\bigvee_{j=1}^{m} (a_i \circ b_j) \right)$, where $a_i \in a \circ Q \cup Q \circ a \cup Q \circ a \circ Q$ and $b_j \in b \circ Q \cup Q \circ b \cup Q \circ b \circ Q$. As Q is commutative $a \circ Q = Q \circ a = Q \circ a \circ Q$ and $b \circ Q = Q \circ b = Q \circ b \circ Q$. Thus, $a_i \circ b_j \in a \circ Q \circ b \circ Q = a \circ b \circ Q$ for all $i = 1, \ldots, n$ and $j = 1, \ldots, m$. Hence, $a_i \circ b_j = a \circ b \circ q \in P$

and then $x \circ y \le \bigvee_{i=1}^{n} \left(\bigvee_{j=1}^{m} (a_i \circ b_j) \right) \in P$. Therefore, $\langle a \rangle \circ \langle b \rangle \subseteq P$ and by the Theorem 2 we have $a \in P$ or $b \in P$.

In what follows, we will introduce the notion of quantale prime. As we know, in ring theory, a quantale is prime iff the ideal generated by 0 is a prime ideal. Then, the Proposition 5 translates this result into quantale environment.

Definition 6 A quantale Q is called prime if given $a, b \in Q$ with $a \ne \bot$ and $b \ne \bot$, there exists $f \in Q$ such that $a \circ f \circ b \ne \bot$.

Proposition 5 *A quantale Q is prime iff $\langle \bot \rangle$ is a prime ideal.*

Proof Suppose Q prime and assume that $I \circ J \subseteq \langle \bot \rangle$, but $I, J \not\subseteq \langle \bot \rangle$, where I, J are ideals of Q. Thus, there exists $i \in I$, $j \in J$ such that $i, j \ne \bot$. As Q is prime, there exists $q \in Q$ such that $i \circ q \circ j \ne \bot$, then we have a contradiction because $i \circ q \circ j \in I \circ J \subseteq \langle \bot \rangle$. Hence, $I \subseteq \langle \bot \rangle$ or $J \subseteq \langle \bot \rangle$. On the other hand, suppose $\langle \bot \rangle$ is a prime ideal of Q. Given $a, b \ne \bot$ in Q, suppose $a \circ q \circ b = \bot$ for all $q \in Q$. Hence, $a \circ Q \circ b \subseteq \langle \bot \rangle$. As $\langle \bot \rangle$ is a prime ideal, then $a \in \langle \bot \rangle$ or $b \in \langle \bot \rangle$, but $a, b \ne \bot$. $\blacksquare$

The Proposition 5 gives a new characterization and opens the investigation on quantales prime.

3   Strong Primeness in Quantales

Strongly prime rings were introduced in 1973, as a prime ring with finite condition in the generalization of results on group rings proved by Lawrence in his master's thesis. In 1975, Lawrence and Handelman [4] came up with properties of those rings and proved important results, for instance all prime rings may be embedded in a strongly prime ring; and all strongly prime rings are nonsingular. After such relevant paper, Olson [8] published a paper about uniform strong primeness and its radical. On the contrary of the concept of strong primeness, Olson proved that the concept of uniform strong primeness is two-sided. In this section we bring this concept to quantales making specific adaptations for this environment. Finally, it is proposed t- and m-systems for quantales, since it gives another characterization of prime and uniformly strongly prime ideal.

Definition 7 Let A be a subset of a quantale Q. The right annihilator of A is defined as $An_r(A) = \{x \in Q : Ax = \langle \bot \rangle\}$. Similarly, we can define the *left annihilator* An_l.

Definition 8 [4] A quantale Q is called right strongly prime if for each $x \in Q - \{\bot\}$ there exists a finite nonempty subset F_x of Q such that $An_r(x \circ F_x) = \langle \bot \rangle$.

Clearly if Q is right strongly prime, then Q is prime. The set F_x is called an insulator of x in Q.

Proposition 6 *If Q is right strongly prime, then every nonzero ideal I of Q contains a finite subset F which has right annihilator zero.*

Proof Suppose Q right strongly prime. Let $x \in I$ and $x \neq \perp$ and $F = x \circ F_x \subseteq I$. Thus, $An_r(F) = \langle \perp \rangle$.

It is clear that every right strongly prime quantale is a prime quantale. It is also possible to define left strongly prime in a similar manner for right strong primeness.

Definition 9 A quantale Q is called uniformly strongly prime (usp) if the same right insulator may be chosen for each nonbottom element.

Proposition 7 *A quantale Q is a right uniformly strongly prime iff there exists a finite subset $F \subseteq Q$ such that for any two nonbottom elements x and y of Q, there exists $f \in F$ such that $x \circ f \circ y \neq \perp$.*

Proof Let Q be uniformly right strongly prime quantale. Hence Q has a uniform right insulator F which is a finite set such that for any element $x \in Q$, $x \circ F$ has no nonbottom right annihilators. Thus, if x and y are any two nonbottom elements in Q, y cannot be in the annihilator of $x \circ F$. Hence there is an $f \in F$ such that $x \circ f \circ y \neq \perp$. For the reverse implication it is easy to see that if the condition is satisfied then for any $x \neq \perp$ in Q, no nonbottom element annihilates $x \circ F$ on the right. Hence Q is uniformly right strongly prime

It is clear that the condition in Proposition 7 is not one-sided; consequently, this condition is also equivalent to uniformly left strongly prime, and we have:

Corollary 3 *Q is uniformly right strongly prime iff Q is uniformly left strongly prime.*

Corollary 4 *A quantale Q is uniformly strongly prime iff there exists a finite subset $F \subseteq Q$ such that $a \circ F \circ b = \perp$ implies $a = \perp$ or $b = \perp$ for all $a, b \in Q$.*

Proof Straightforward.

Definition 10 An ideal $P \neq \langle \perp \rangle$ of a quantale Q is called uniformly strongly prime (usp) ideal if there exists a finite subset $F \subseteq Q$ such that $a \circ F \circ b \subseteq P$ implies $a \in P$ or $b \in P$.

Proposition 8 *An ideal I of a quantale Q is a usp ideal iff there exists a finite subset $F \subseteq Q$ such that for any two elements $a, b \in Q \setminus I$(complement of I in Q), there exists $f \in F$ such that $x \circ f \circ y \notin I$.*

Proof Suppose I a usp ideal of Q. If $a \notin I$ and $b \notin I$ by Definition 10 $a \circ F \circ b$ is not a subset of I. Hence, there exists $f \in F$ such that $a \circ f \circ b \notin I$. For the converse, note that by hypothesis it is impossible to have $a \circ F \circ b \subseteq I$ and $a \notin I$ and $b \notin I$.

Subsequently we introduce the t-/m-systems. They will give us another characterization of prime and usp ideals.

Definition 11 A subset M of a quantale Q is called an *m-system* if for any two elements $x, y \in M$ there exists $q \in Q$ such that $x \circ q \circ y \in M$.

Definition 12 A subset T of a quantale Q is called a *t-system* if there exists a finite subset $F \subseteq Q$ such that for any two elements $x, y \in T$ there exists $f \in F$ such that $x \circ f \circ y \in T$.

Proposition 9 *I is a prime ideal of a quantale Q iff $Q \setminus I$ (the complement of I in Q) is an m-system.*

Proof Suppose I a prime ideal. If $a, b \in R \setminus I$, then $a, b \notin I$. By Proposition 2 the subset $a \circ Q \circ b$ is not a subset of I. Thus, there exists $q \in Q$ such that $a \circ q \circ b \notin I$. For the converse, let $a, b \in Q$ such that $a \circ Q \circ b \subseteq I$, if a and b not in I, then $a, b \in Q \setminus I$. By hypothesis there exists $q \in Q$ such that $a \circ q \circ b \in Q \setminus I$, but $a \circ Q \circ b \subseteq I$.

Proposition 10 *An ideal I is a usp ideal of a quantale Q iff $Q \setminus I$ (the complement of I in R) is a t-system.*

Proof This Proposition is similar to Proposition 8.

At the end of Sect. 2 we introduced a quantale prime where it was proposed a right/left strongly quantale prime. Lawrence and Handelman proved that all prime rings may be embedded in a strongly prime ring. Then, a question arises: based on their studies, may we have the similar result in quantales?

4 Fuzzy Prime and Fuzzy usp Ideals in Quantales

In 1965, Zadeh [9] introduced fuzzy sets and in 1971, Rosenfeld [10] introduced fuzzy sets in the realm of group theory and formulated the concept of fuzzy subgroups of a group. Since then, many researchers have been engaged in extending the concepts/results of abstract algebra to the broader framework of the fuzzy setting. Thus, in 1982, Liu [11] defined and studied fuzzy subrings as well as fuzzy ideals. Subsequently, Liu [12], Mukherjee and Sen [13], Swamy and Swamy [14], and Zhang Yue [15] fuzzified certain standard concepts/results on rings and ideals. For example: Mukherjee was the first to study the notion of prime ideal in a fuzzy setting. Those studies were investigated by Kumar in [16] and [17], where the notion of a nil radical and semiprimeness were introduced.

After Mukherjee's definition of prime ideals in the fuzzy setting, many investigations extended crisp results to fuzzy setting. But Mukherjee's definition was not appropriate to deal with noncommutative rings. In 2012, Navarro, Cortadellas and Lobillo [3] drew attention to this specific problem. They proposed a new definition of primeness for fuzzy ideals for noncommutative rings holding the idea of "fuzzification" of primeness introduced by Kumbhojkar and Bapat [18, 19] to commutative rings, which is coherent with α-cuts. This section introduces the first version of fuzzy prime ideals and fuzzy uniformly strongly prime ideals in quantales compatible with the ideas developed in [3] i.e. a fuzzy concept on membership function is defined and after that it is proved a coherency with α-cuts. It is also proved that every fuzzy completely prime ideal is a fuzzy prime ideal but the converse is not true in noncommutative quantale according to Proposition 3.

By a fuzzy subset we mean the classical concept defined by Zadeh [9], i.e., a fuzzy set over a base set X is a set map $\mu : X \longrightarrow [0, 1]$. The intersection and union of fuzzy sets are given by the point-by-point infimum and supremum. We shall use the symbols $\wedge$ and $\vee$ for denoting the infimum and supremum of a collection of real numbers. Hence, $\bigvee A$ is the supremum of a set A and $\bigwedge A$ is the infimum of a set A. Again, $x \circ A$ denotes the set $\{x \circ a : a \in A\}$ and $x \circ A \circ y = \{x \circ a \circ y : a \in A\}$.

Definition 13 [2] Let Q be a quantale. A fuzzy subset I of Q is called a fuzzy ideal of Q if it satisfies the following conditions for $x, y \in Q$:

(1) if $x \leq y$, then $I(x) \leq I(y)$;
(2) $I(x \vee y) \geq I(x) \wedge I(y)$;
(3) $I(x \circ y) \geq I(x) \vee I(y)$.

From (1) and (2) it follows that $I(x \vee y) = I(x) \wedge I(y)$. Thus, a fuzzy subset I is a fuzzy ideal of Q iff $I(x \vee y) = I(x) \wedge I(y)$ and $I(x \circ y) \geq I(x) \vee I(y)$.

Let μ be a fuzzy subset of X and let $\alpha \in [0, 1]$. Then the set $\{x \in X : \mu(x) \geq \alpha\}$ is called the α-cut. Clearly, if $t > s$, then $\mu_t \subseteq \mu_s$. Again, it is proved in [2] that I is a fuzzy ideal of Q iff I_α is an ideal of Q for all $\alpha \in (I(\top), 1]$.

Definition 14 A non-constant fuzzy ideal $P : Q \longrightarrow [0, 1]$ is a fuzzy prime ideal of Q if for any $x, y \in Q$, $\bigwedge P(x \circ Q \circ y) = P(x) \vee P(y)$.

Definition 15 A non-constant fuzzy ideal $P : Q \longrightarrow [0, 1]$ is said to be fuzzy completely prime (fcp) ideal of Q if for any $x, y \in Q$, $P(x \circ y) = P(x)$ or $P(x \circ y) = P(y)$.

Definition 16 A non-constant fuzzy ideal $I : Q \longrightarrow [0, 1]$ is said to be fuzzy uniformly strongly prime (fusp) ideal if there exists a finite subset $F \subseteq Q$ such that $\bigwedge I(x \circ F \circ y) = I(x) \vee I(y)$, for any $x, y \in Q$. The subset F is called insulator of I in Q.

The following Proposition says that the definition of fuzzy uniformly strongly prime is coherent with the α-cuts.

Proposition 11 *I is a fuzzy prime ideal of Q iff I_α is a prime ideal of Q for all $\alpha \in (I(\top), 1]$.*

Proof Suppose I a fuzzy ideal of Q. Let $x, y \in Q$ such that $x \circ Q \circ y \subseteq I_\alpha$. Thus, $x \circ q \circ y \in I_\alpha$ for all $q \in Q$. As I is a fuzzy prime, then we have $I(x) \vee I(y) = \bigwedge I(x \circ Q \circ y) \geq \alpha$ hence $I(x) \geq \alpha$ or $I(y) \geq \alpha$ i.e. $x \in I_\alpha$ or $y \in I_\alpha$. Thus, by Theorem 2 I_α is a prime ideal. On the other hand, suppose I_α prime ideal of Q for all $\alpha \in (I(\top), 1]$ and $\bigwedge I(x \circ Q \circ y) > I(x) \vee I(y)$. Let $t = \bigwedge I(x \circ Q \circ y)$, and thus $t > I(x) \vee I(y)$ and $x, y \notin I_t$, but this is a contradiction because $I(x \circ q \circ y) \geq t$ for all $q \in Q$ i.e. $x \circ Q \circ y \subseteq I_t$, as I_t is a prime ideal then $x \in I_t$ or $y \in I_t$. Therefore, I is a fuzzy prime ideal.

Proposition 12 *[2] For a fuzzy ideal P in Q the following statements are equivalent:*

(1) P is fuzzy completely prime ideal;
(2) $I(x \circ y) = I(x) \vee I(y)$ for all $x, y \in Q$;
(3) I_α is completely prime ideal of Q for all $\alpha \in (I(\top), 1]$.

Proposition 13 *If P is a completely fuzzy prime ideal of Q, then P is a fuzzy prime ideal of Q.*

Proof Use Proposition 11 and Proposition 12.

Proposition 14 *In a commutative quantale an ideal is completely fuzzy prime iff is fuzzy prime.*

Proof Use Propositions 4, 11 and 12.

Theorem 5 *For a fuzzy ideal P in Q the following statements are equivalent:*

(1) P is a fuzzy prime ideal;
(2) Given $x, y \in Q$ and J a fuzzy ideal of Q we have: $J(x \circ r \circ y) \leq P(x \circ r \circ y)$ for all $r \in Q$ implies $J(x) \leq P(x)$ or $J(y) \leq P(y)$.

Proof $(1) \Rightarrow (2)$, suppose P fuzzy prime ideal, if $J(x \circ q \circ y) \leq P(x \circ q \circ y)$ for all $q \in Q$, then $\bigwedge J(x \circ Q \circ y) \leq \bigwedge P(x \circ Q \circ y)$. As J is a fuzzy ideal, by Definition 13 (3) we have $J(x \circ r \circ y) \geq J(x) \vee J(r) \vee J(y) \geq J(x) \vee J(y)$, hence $\bigwedge J(x \circ Q \circ y) \geq J(x) \vee J(y)$. Thus, $J(x) \vee J(y) \leq \bigwedge J(x \circ Q \circ y) \leq \bigwedge P(x \circ Q \circ y) = P(x) \vee P(y)$. Hence, $J(x) \vee J(y) \leq P(x) \vee P(y)$. Therefore, $J(x) \leq P(x)$ or $J(y) \leq P(y)$. For $(2) \Rightarrow (1)$, suppose that $\bigwedge P(x \circ Q \circ y) > P(x) \vee P(y)$ for some $x, y \in Q$. Then there exists $t \in (0, 1)$ such that $\bigwedge P(x \circ Q \circ y) > t > P(x) \vee P(y)$. Now, define the ideal $I : Q \longrightarrow [0, 1]$ given by:
$$I(z) = \begin{cases} P(z), & if\ P(z) \geq t \\ t, & otherwise \end{cases}$$
This is a fuzzy ideal with $t < I(x \circ q \circ y) = P(x \circ q \circ y)$ for all $q \in Q$, but $t = I(x) = I(y) > P(x) \vee P(y)$.

5 Final Remarks

As we known, for mathematics prime ideals have developed an important role in ring theory and have attracted the attention of some researchers in the investigation of quantales. As prime ideals are structural pieces of a ring it is relevant to study its concept in order to establish a well-founded quantale theory. With this in mind, it is necessary to investigate primeness over arbitrary quantales, i.e. commutative and noncommutative setting. Therefore, this paper invites the reader to think about primeness in quantales bringing a new perspective for prime ideals.

Acknowledgements The authors would like to thank UESB (Southwest Bahia State University) and UFRN (Federal University of Rio Grande do Norte) for their financial support. This research was partially supported by the Brazilian Research Council (CNPq) under the process 306876/2012-4.

References

1. Yang, L., Luoshan, X.: Roughness in quantales. Inf. Sci. **220**, 568–579 (2013)
2. Luo, Q., Wang, G.: Roughness and fuzziness in quantales. Inf. Sci. **271**, 14–30 (2014)
3. Navarro, G., Cortadellas, O., Lobillo, F.J.: Prime fuzzy ideals over noncommutative rings. Fuzzy Sets Syst. **199**(0):108–120 (2012)
4. Handelman, D., Lawrence, J.: Strongly prime rings. Trans. Am. Math. Soc. **211**, 209–223 (1975)
5. Bergamaschi, F.B., Santiago, R.H.N.: Strongly prime fuzzy ideals over noncommutative rings. In: 2013 IEEE International Conference on Fuzzy Systems (FUZZ), pp. 1–5 (2013)
6. Bergamaschi, F.B., Santiago, R.H.N.: A fuzzy version of uniformly strongly prime ideals. In: 2014 IEEE Conference on Norbert Wiener in the 21st Century (21CW), pp. 1–6. IEEE (2014)
7. Rosenthal, K.I.: Quantales and Their Applications, vol. 234. Longman Scientific & Technical Harlow (1990)
8. Olson, D.M.: A uniformly strongly prime radical. J. Austral. Math. Soc. (Ser. A) **43**, 95–102 (1987)
9. Zadeh, L.A.: Fuzzy sets. Inf. Control **8**(3), 338–353 (1965)
10. Rosenfeld, A.: Fuzzy groups. J. Math. Anal. Appl. **35**(3), 512–517 (1971)
11. Wang jin Liu: Fuzzy invariant subgroups and fuzzy ideals. Fuzzy Sets Syst. **8**(2), 133–139 (1982)
12. Wang-jin, L.: Operations on fuzzy ideals. Fuzzy Sets Syst. **11**(13), 19–29 (1983)
13. Mukherjee, T.K., Sen, M.K.: On fuzzy ideals of a ring. Fuzzy Sets Syst. **21**(1), 99–104 (1987)
14. Swamy, U.M., Swamy, K.L.N.: Fuzzy prime ideals of rings. J. Math. Anal. Appl. **134**(1), 94–103 (1988)
15. Yue, Z.: Prime l-fuzzy ideals and primary l-fuzzy ideals. Fuzzy Sets Syst. **27**(3), 345–350 (1988)
16. Kumar, R.: Fuzzy semiprimary ideals of rings. Fuzzy Sets Syst. **42**(2), 263–272 (1991)
17. Kumar, R.: Fuzzy nil radicals and fuzzy primary ideals. Fuzzy Sets Syst. **43**(1), 81–93 (1991)
18. Kumbhojkar, H.V., Bapat, M.S.: Not-so-fuzzy fuzzy ideals. Fuzzy Sets Syst. **37**(2), 237–243 (1990)
19. Kumbhojkar, H.V., Bapat, M.S.: On prime and primary fuzzy ideals and their radicals. Fuzzy Sets Syst. **53**(2), 203–216 (1993)

A Short Introduction to Collective Defense in Weighted Abstract Argumentation Problems

Stefano Bistarelli

Abstract In this paper we review the basic ideas of weighted argumentation framework that we exploited in the last years. In particular we show how to deal with Argumentation Frameworks with weight on attacks. Such extension clearly add more information: from one side we can compare attacks measuring the added weight on the attack itself; from the other side we can aggregate attacks together using a semiring combination operator able to compute the quantitative strength of the synergy. The additional weight associated to attack relations foster a new definition of defense able to compare the weights associated to the attacks, assuring that the defense happen when it is stronger than the attack.

1 Introduction and Preliminaries

The aim of this work is to give a short description of weighted argumentation as introduced in [10–14]. An *Abstract Argumentation Framework* (*AAF*) [19] is represented by a pair $\langle \mathcal{A}_{rgs}, R \rangle$ consisting of a set of arguments and a binary relation of attack defined among them. Given such a framework, it is possible to examine the question on which set(s) (extension(s)) of arguments can be accepted, hence collectively surviving the conflict defined by R. Answering this question corresponds to defining an argumentation semantics.

A very simple example of AAF is $\langle \{a, b\}, \{R(a, b), R(b, a)\} \rangle$, where two arguments a and b attack each other. In this case, each of the two positions represented by either $\{a\}$ or $\{b\}$ can be intuitively valid, since no additional information is provided on which of the two attacks prevails.

This example gives motivations for introducing a weighted notion of AF. Indeed, depending on which attack $a \to b$ or $b \to a$ is stronger leads one of the two arguments to prevail. For instance, in case the attack $R(a, b)$ is stronger than (or preferred to) $R(b, a)$, taking the position defined by a may result in a better choice for an intel-

S. Bistarelli (✉)
University of Perugia, Perugia, Italy
e-mail: stefano.bistarelli@unipg.it

© Springer Nature Switzerland AG 2020
M. Ceberio and V. Kreinovich (eds.), *Decision Making under Constraints*,
Studies in Systems, Decision and Control 276,
https://doi.org/10.1007/978-3-030-40814-5_5

ligent agent, since it can be defended better. The first step to deal with weight is then a new notion of weighted defence (in classical argumentation, an argument a defends an argument c from b, if $a \rightarrow b \rightarrow c$). In weighted argumentation an argument a is defending c only if the weight of the attack from a to b is stronger than the attack from b to c. The following definition generalize the defense to a set of arguments.

Definition 1 (*Defence* ($\mathbb{D}_0$) [19]) An argument b is defended by a set $\mathcal{B} \subseteq \mathcal{A}_{rgs}$ (or $\mathcal{B}$ defends b) iff for any argument $a \in \mathcal{A}_{rgs}$, if $R(a, b)$ then $\exists b \in \mathcal{B}$ s.t., $R(b, a)$.

An admissible set of arguments must be a conflict-free set that defends all its elements. Formally:

Definition 2 (*Admissible sets* [19]) A conflict-free set $\mathcal{B} \subseteq \mathcal{A}_{rgs}$ is admissible iff each argument in $\mathcal{B}$ is defended by $\mathcal{B}$.

Three classical semantics refining admissibility are defined in the following definitions:

Definition 3 (*Complete, Preferred and Stable semantics* [19]) An admissible set $\mathcal{B} \subseteq \mathcal{A}_{rgs}$ is a complete extension iff each argument which is defended by $\mathcal{B}$ is in $\mathcal{B}$. A preferred extension is a maximal (w.r.t. set inclusion) admissible subset of $\mathcal{A}_{rgs}$. A conflict-free set $\mathcal{B} \subseteq \mathcal{A}_{rgs}$ is a stable extension iff for each argument which is not in $\mathcal{B}$, there exists an argument in $\mathcal{B}$ that attacks it.

2 Weighted Abstract AFs

To have a general and formal representation of weights and operations on them (i.e., aggregation and preference), we use a parametric algebraic framework based on c-semirings [5, 7].[1] Hence, it is possible to consider different metrics within the same computational framework, as fuzzy or probabilistic scores, and model different kinds of AAFs in the literature (see Sect. 4).

In Fig. 1 we provide an example of a weighted AAF describing the $WAAF_{\mathbb{S}}$ defined by $\mathcal{A}_{rgs} = \{a, b, c, d, e\}$, $R = \{(a, b), (c, b), (c, d), (d, c), (d, e), (e, e)\}$, with $W(a, b) = 7, W(c, b) = 8, W(c, d) = 9, W(d, c) = 8, W(d, e) = 5$, $W(e, e) = 6$, and $\mathbb{S} = \langle \mathbb{R}^+ \cup \{\infty\}, \min, +, \infty, 0 \rangle$ (i.e., the weighted semiring).

Therefore, each attack is associated with a semiring value that represents the "strength" of an attack between two arguments. We can consider the weights in Fig. 1 as supports to the associated attack, as similarly suggested in [20]. A semiring value equal to the top element of the c-semiring $\top$ (e.g., 0 for the weighted semiring) represents a no-attack relation between two arguments: for instance, $(a, c) \notin R$ in

---

[1]In practice, c-semirings are *commutative* ($\otimes$ is commutative) and *idempotent* semirings (i.e., $\oplus$ is idempotent), where $\oplus$ defines a complete lattice: every subset of elements have a *least upper bound*, or *lub*, and a *greatest lower bound*, or *glb*. In fact, c-semirings are semirings where $\oplus$ is used as a preference operator, while $\otimes$ is used to compose preference-values together.

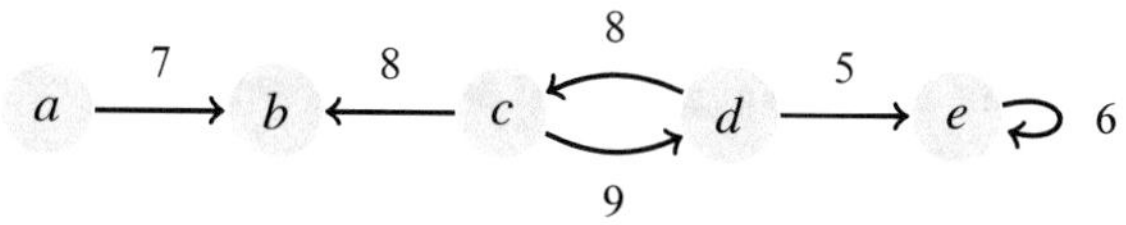

Fig. 1 An example of
WAAF

Fig. 1 corresponds to $W(a, c) = 0$. Note that, whenever there is an attack between two arguments, its weight is different from $\top$: for example, $W(a, b) = 7$ in Fig. 1. On the other side, the bottom element, i.e., $\bot$ (e.g., ∞ for the weighted semiring), represents the strongest attack possible.

In Definition 4 we define the attack strength for a set of arguments that attacks an argument, a different set of arguments, or an argument that attacks a set of arguments; the former and the latter are what we need to define w-defence. In the following, we will use $\bigotimes$ to indicate the $\otimes$ operator of the c-semiring $\mathbb{S}$ on a set of values:

Definition 4 (*Attacks to/from sets of arguments*) Given a $WAAF_{\mathbb{S}}$, $WF = \langle \mathcal{A}_{rgs}, R, W, \mathbb{S} \rangle$,

- a set of arguments $\mathcal{B}$ attacks an argument a with a weight of $k \in S$ if

$$W(\mathcal{B}, a) = \bigotimes_{b \in \mathcal{B}} W(b, a) = k$$

- an argument a attacks a set of arguments $\mathcal{B}$ with a weight of $k \in S$ if

$$W(a, \mathcal{B}) = \bigotimes_{b \in \mathcal{B}} W(a, b) = k$$

- a set of arguments $\mathcal{B}$ attacks a set of arguments $\mathcal{D}$ with a weight $k \in S$ if

$$W(\mathcal{B}, \mathcal{D}) = \bigotimes_{b \in \mathcal{B}, d \in \mathcal{D}} W(b, d) = k$$

For example, looking at Fig. 1 we have that $W(\{a, c\}, b) = 15$, $W(c, \{b, d\}) = 17$, and $W(\{a, c\}, \{b, d\}) = 24$.

Considering the possibility to combine together more attack has been already highlighted important by Nielsen and Parsons [28]. Here we extend the synergic effect of several attacks to a single argument by also considering the combined effect of several attacks from one argument to a set of arguments. Also, since the attacks are equipped with a numeric strength, also such numeric value need to be computed, and this is done using the semiring times operator.

We are now ready to define our version of weighted defence, i.e., w-defence:

Definition 5 (*w-defence* ($\mathbb{D}_w$)) Given a $WAAF_{\mathbb{S}}$, $WF = \langle \mathcal{A}_{rgs}, R, W, S \rangle$, $\mathcal{B} \subseteq \mathcal{A}_{rgs}$ w-defends $b \in \mathcal{A}_{rgs}$ iff $\forall a \in \mathcal{A}_{rgs}$ such that $R(a, b)$, we have that

$$W(a, \mathcal{B} \cup \{b\}) \geq_{\mathbb{S}} W(\mathcal{B}, a)$$

As previously advanced, a set $\mathcal{B} \subseteq \mathcal{A}_{rgs}$ w-defends an argument b from a, if the $\otimes$ of all attack weights from $\mathcal{B}$ to a is worse[2] (w.r.t. $\leq_S$) than the $\otimes$ of the attacks from a to $\mathcal{B} \cup \{b\}$.

For example, the set $\{c\}$ in Fig. 1 defends c from d because $W(d, \{c\}) \geq_S W(\{c\}, d)$, i.e., $(8 \leq 9)$. On the other hand, $\{d\}$ in Fig. 1 does not w-defend d because $W(c, \{d\}) \not\geq_S W(\{d\}, c)$.

In our proposal, an extension $\mathcal{B} \subseteq \mathcal{A}_{rgs}$ defends an argument $b \in \mathcal{A}_{rgs}$ from $a \in \mathcal{A}_{rgs}$, if the composition (a parametric $\otimes$ operation from a c-semiring structure [5]) of all the attack weights from $\mathcal{B}$ to a is stronger than the composition of all the attacks from a to $\mathcal{B}$. Differently from [18], where the arithmetic sum of all attack weights from $\mathcal{B}$ to a needs to be only stronger than the attack from a to b, we also consider the set of attacks from a to $\mathcal{B}$. Therefore, both our proposal and the one given by [18] suggest a collective defence from $\mathcal{B}$ to a, but, differently, in this paper we consider the group of attacks from a to $\mathcal{B}$ as a single entity, i.e., with a single global weight. We believe such a choice provides a more coherent view: in the literature, defence is usually checked by considering all the counter-attacks from a set $\mathcal{B}$ to a (e.g., in order to satisfy admissibility), but each attack from a to $\mathcal{B}$ is treated separately (however, in case of fuzzy aggregation of weights, the two approaches are equivalent). Our intent is to normalize such dis-homogeneity.

As defined, w-defence implies the classical Dung's defence in Definition 1:

Proposition 1 ($\mathbb{D}_w \Rightarrow \mathbb{D}_0$ [12]) *Given a WAAF$_S$, WF $= \langle \mathcal{A}_{rgs}, R, W, S \rangle$, a subset of arguments $\mathcal{B}$, and $b \in \mathcal{A}_{rgs}$, "$\mathcal{B}$ w-defends b" (Definition 5) $\Rightarrow$ "$\mathcal{B}$ defends b" (Definition 1) in the corresponding not-weighted $\langle \mathcal{A}_{rgs}, R \rangle$.*

Moreover, the following proposition equates defence and w-defence in case we adopt the boolean c-semiring:

Proposition 2 ([12]) *Given a WAAF$_S$, WF $= \langle \mathcal{A}_{rgs}, R, W, \mathbb{S} \rangle$, where $\mathbb{S} = \langle \{true, false\}, \vee, \wedge, false, true \rangle$ (i.e., the boolean semiring), "$\mathcal{B}$ w-defends a" $\Longleftrightarrow$ "$\mathcal{B}$ defends a".*

3   Implementation

We have implemented our weighted argomentation framework in *ConArg*[3] [6], which is a tool that exploits *Gecode*[4] (a constraint-programming library) to solve several problems related to Argumentation.

---

[2]Note that, when considering the partial order of a generic semiring, we will often use "worse" or "better" because "greater" or "lesser" would be misleading: in the weighted semiring, $7 \leq_S 3$, i.e., lesser means better.

[3]http://www.dmi.unipg.it/conarg/.

[4]http://www.gecode.org.

4 Related Work and Comparison

Two of the most related definitions of weighted defence (i.e., Definition 5) are presented in [18, 26]. In [26] attacks are relatively ordered by their force, i.e., $R(a, b) \gg R(b, a)$ means that the former attack is stronger than the latter (vice-versa, a weaker attack). Equivalent and incomparable classes are considered as well, i.e., $R(a, b) \approx R(b, a)$ and $R(a, b)?R(b, a)$, respectively. This is accordingly reflected by the defence definition, where considering $R(a, b)$ and $R(c, a)$ we can have that c is a *strong*, *weak*, *normal*, or *unqualified* defender of b. Therefore, an argument b is defended by $\mathcal{B}$ if, and only if, for any argument a such that $R(a, b)$, there is an argument $c \in \mathcal{B}$ such that $R(c, a)$, and according to the desired defence strength, $R(c, a) \gg R(a, b)$, $R(c, a) \ll R(a, b)$, $R(c, a) \approx R(a, b)$, and $R(c, a)?R(a, b)$. For instance, when requiring a level $[\gg, \approx]$, for each attacker a of b there must must be either a strong or a normal defender $c \in \mathcal{B}$. In Definition 6 we exactly rephrase such defence by modelling the total order defined by $[\gg, \approx]$ with a c-semiring $\mathbb{S}$:

Definition 6 (*Defence* $\mathbb{D}_1$ [26]) Given $WF = \langle \mathcal{A}_{rgs}, R, W, \mathbb{S} \rangle$, $a, b, c \in \mathcal{A}_{rgs}$, $\mathcal{B} \subseteq \mathcal{A}_{rgs}$, then b is defended by $\mathcal{B}$ if $\forall R(a, b), \exists c \in \mathcal{B}$ s.t. $W(a, b) \geq_{\mathbb{S}} W(c, a)$.

In [18] the authors define $\sigma^{\boxtimes}$-extensions, where σ is one of the given semantics (*e.g.*, admissible), and $\boxtimes$ is an *aggregation function* ($\otimes$ in a c-semiring). $\boxtimes$ needs to satisfy non-decreasingness, minimality, and identity[5]: two examples in the paper are the arithmetic sum and max. Even the notion of defence is refined: in Definition 7 we cast it in the same semiring-based framework.

Definition 7 (*Defence* $\mathbb{D}_2$ [18]) Given $WF = \langle \mathcal{A}_{rgs}, R, W, \mathbb{S} \rangle$, an argument b is defended by a subset of arguments $\mathcal{B}$ if $\forall a \in \mathcal{A}_{rgs}$ s.t. $R(a, b)$, we have that $W(a, b) \geq_{\mathbb{S}} W(\mathcal{B}, a)$.

Thus, an argument b is $\boxtimes$-acceptable if for each attack from a against b, the aggregated weight of the collective defence of b is greater than $W(a, b)$, according to $\mathbb{D}_2$. Such phrasing of defence is also equivalent to what presented in [13].

By using the same semiring-based framework, it is now possible to relate such notions of defence together (we remind that $\mathbb{D}_w$ stands for w-defence).

Theorem 1 ([12])

- $\mathbb{D}_w \Rightarrow \mathbb{D}_2$;
- $\mathbb{D}_1 \Rightarrow \mathbb{D}_2$;
- *If* $\mathbb{S} = \langle [0, 1], \max, \min, 0, 1 \rangle$, *then* $\mathbb{D}_w \Rightarrow \mathbb{D}_1 \Leftrightarrow \mathbb{D}_2$;
- *If* $\mathbb{S} = \langle \{true, false\}, \vee, \wedge, false, true \rangle$, *then* $\mathbb{D}_w \Leftrightarrow \mathbb{D}_0 \Leftrightarrow \mathbb{D}_1 \Leftrightarrow \mathbb{D}_2$.

An example on how $\mathbb{D}_w$, $\mathbb{D}_1$, and $\mathbb{D}_2$ differently work is provided in Fig. 2. We read this example by considering the weighted c-semiring, i.e., $\mathbb{S} = \langle \mathbb{R}^+ \cup \{+\infty\}, \min, +, +\infty, 0 \rangle$. Argument b is defended by $\mathcal{B} = \{b, c, d, e\}$ according

---

[5]Such properties are satisfied by a c-semiring.

Fig. 2 An example of
WAAF where $\{b, c, d, e\}$ are
defended by $\mathcal{B}$ according to
$\mathbb{D}_2$ (using the weighted
semiring)

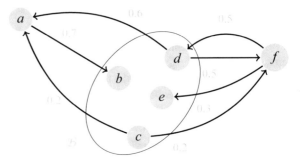

Fig. 3 An example of AAF

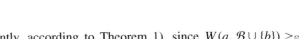

to $\mathbb{D}_w$ and $\mathbb{D}_2$ (consequently, according to Theorem 1), since $W(a, \mathcal{B} \cup \{b\}) \geq_S$
$W(\mathcal{B}, a)$ $(0.7 \leq 0.8)$. It is not defended according to $\mathbb{D}_1$, since $W(d, a) \geq_S W(a, b)$
$(0.6 \leq 0.7)$ and $W(c, a) \geq_S W(a, b)$ $(0.2 \leq 0.7)$.

On the other hand, considering the attacks from f to $\mathcal{B}$ instead, $\mathbb{D}_w$ does not hold:
$W(f, d) \otimes W(f, e) \not\geq_S W(d, f) \otimes W(c, f)$ (i.e., $0.8 \not\leq 0.7$). However, $\mathbb{D}_2$ holds
because $W(f, d) \geq_S W(\{d, c\}, f)$ (i.e., $0.5 \leq 0.7$) and $W(f, e) \geq_S W(\{d, c\}, f)$
(i.e., $0.3 \leq 0.7$). Therefore, considering the WAAF in Fig. 2, $\mathcal{B}$ defends itself from
a and f according to only $\mathbb{D}_2$ ($\mathbb{D}_w$ and $\mathbb{D}_1$ do not hold).

Reading the same example in Fig. 2 with $\mathbb{S} = \langle [0, 1], \max, \min, 0, 1 \rangle$ instead, $\mathbb{D}_2$
collapses to $\mathbb{D}_1$ and $\mathbb{D}_1$ does not hold due to $R(a, b)$. Moreover, since $\mathbb{D}_1$ is not valid
then $\mathbb{D}_w$ cannot hold as well.

This example introduces us to relating admissible sets (at the core of all the
extensions in [19]) using $\mathbb{D}_w$, $\mathbb{D}_1$, and $\mathbb{D}_2$. We focus on this semantics because it is
at the core of the other ones proposed by [19], explicitly (i.e., complete, preferred,
grounded), or implicitly (i.e., stable). We respectively call adm_1 and adm_2 the set
of admissible sets using $\mathbb{D}_1$ and $\mathbb{D}_2$, adm_w is our proposal, and adm_0 is the classical
definition [19].

Theorem 2 ([12]) *Given $WF = \langle \mathcal{A}_{rgs}, R, W, \mathbb{S} \rangle$ where $\mathbb{S} = \langle [0, 1], \max, \min, 0, 1 \rangle$,
then $adm_w(WF) = adm_1(WF) = adm_2(WF) \subseteq adm_0(WF)$.*

Considering the framework in Fig. 1 as unweighted (i.e., the one in Fig. 3), Dung's
admissible sets are: $\{a\}$, $\{c\}$, $\{d\}$, $\{a, c\}$, $\{a, d\}$. admissible$-w$ sets are $\{a\}$, $\{c\}$, and
$\{a, c\}$ instead: $\{a\}$ because is not attacked by any other argument, $\{c\}$ and $\{a, c\}$
because they both w-defends c from the attack performed by d, i.e., $W(d, c) \geq_S$
$W(c, d)$ (i.e., $8 \leq 9$). For instance $\{d\}$ is not admissible$_w$ because it is not able to w-
defend itself from the attack of c. For the same reason, $\{a, d\}$ is not w-admissible. As
we can see from this example, w-defence restrain Dung's defence, and, accordingly
the number of sets satisfying admissibility: in this case we drop $\{d\}$ and $\{a, d\}$ w.r.t.
[19].

The classical inclusion-relations [19] among the newly-defined weighted semantics (annotated by the $w-$) are still valid:

Theorem 3 (semantics inclusions [12]) *Given any* $\langle \mathcal{A}_{rgs}, R, W, \mathbb{S} \rangle$, *with* $\mathbb{S} = \langle S, \oplus, \otimes, \bot, \top \rangle$, *and* $\alpha, \gamma \in S$,

1. *each* $w-$*admissible set is also* $w-$*conflict-free.*
2. *each* $w-$*complete extension is also a* $w-$*admissible set.*
3. *each* $w-$*preferred extension is also* $w-$*complete.*
4. *each* $w-$*stable extension is also* $w-$*preferred.*

Besides the technical comparison between the most related definitions of weighted defence [18, 26] also other frameworks in the literature take into account weights or preference-values on either arguments or attacks.

In [3], AAFs have been extended to *Value-based* AAFs (*VAFs*). A VAF is a five-tuple $\langle \mathcal{A}_{rgs}, R, V, val, P \rangle$, where $\mathcal{A}_{rgs}$ is a finite set of arguments, R is an irreflexive binary relation on A (i.e., $\langle \mathcal{A}_{rgs}, R \rangle$ is a standard AAF), V is a non-empty set of values, val is a function which maps from elements of A to elements of V, and P is the set of possible audiences (i.e., total orders on V). We say that an argument a relates to value v if accepting A promotes or defends v: the value in question is given by $val(a)$. For every $a \in \mathcal{A}_{rgs}$, $val(a) \in V$. When the VAF is considered by a particular audience, the value ordering is fixed.

A *Preference-based* argumentation AAF [2] is a triplet $\langle \mathcal{A}_{rgs}, R, Pref \rangle$ where *Pref* is a partial pre-ordering (reflexive and transitive binary relation) on $\mathcal{A}_{rgs} \times \mathcal{A}_{rgs}$. The notion of defence changes accordingly: let a and b be two arguments, b attacks a iff $R(b, a)$ and not $a > b$, i.e., a is not preferred in the partial pre-ordering.

In [27], the author extends Dung's theory of argumentation to integrate a meta-level argumentation concerning preferences. Dung's level of abstraction is preserved, so that arguments expressing preferences are distinguished by being the source of a second attack relation. This abstractly characterises the application of preferences by attacking attacks between the arguments that are subject to preference claims.

Hence, the first three presented references "weigh" arguments instead of attacks, and, moreover, the proposed frameworks are qualitatively-oriented instead of quantitatively-oriented (as in our proposal). To comment on the first issue, we remind that it can be possible to aggregate the weights on the attacks in order to obtain a problem with preferences/scores: examples of such approaches are explained in [13, 24, 27]. Note that the semiring operator $\oplus$ can represent both a partial [2] and a total [3] order among the arguments.

A quantitative study is proposed in [24], where the authors define *Social Abstract Argumentation Frameworks*, which basically associate positive and negative votes to each argument. Afterwards, it is defined how to aggregate these votes together, and how to associate it with an unique social model. This framework has been extend in [21] by considering weights on attacks as well.

An approach to use argumentation as voting methods is instead used in [4]. Here extensions represent a non-conflicting committee to be elected.

In [17], the authors review the works in [2, 16, 20, 23], focusing on how to relate preference-values and weights, on either arguments or attacks. In [13], if $R(a, b)$ and $R(b, c)$, a defends c if $W(a, b)$ is worse than $W(b, c)$ (as in [23]), thus the defence is not collective as instead in [18] and this paper, and the attack is not collective as in this work. In [23] the difference between the weight associated with a is related to both the weights of b and c, with the purpose to check how much a defends b (thus obtaining "*varied-strength defeat relations*").

The two principles in [22] are, (i) having fewer attackers is better than having more, and (ii) having more defenders is better than having fewer. The result is the definition of a graded defence $d_{m,n}(\mathcal{E})$, which defines different levels of defence-strength: if $d_{m,n}(\mathcal{E})$ holds, $\mathcal{E}$ is a set of arguments for which each $a \in \mathcal{E}$ does not have at least m attackers that are not counter-attacked by at least n arguments in $\mathcal{E}$. Hence, the notions of defence in [1, 22, 23] follow a different approach and cannot straightforwardly be represented by our framework.

One of the main advantages of the general semiring-based framework proposed here is to be capable of modelling several different WAAFs. This results into a comprehensive computational-model for Weighted Abstract Argumentation. Also, even classical Dung's semantics can be modelled in the same framework.

In the following of this section we report different quantitatively-oriented WAAFs in literature [20, 25, 29], which are all encompassed in our general approach. For instance, an argument can be seen as a chain of possible events that makes a hypothesis true [25]. The credibility of a hypothesis can then be measured by the total probability that it is supported by arguments. To solve this problem we can use the probabilistic semiring $\langle [0..1], \max, \times, 0, 1 \rangle$, where the arithmetic multiplication (i.e., $\times$) is used to compose the probability values together (assuming that the probabilities being composed are independent). In [25] the authors associate probabilities with arguments and defeats. Then, they compute the likelihood of some set of arguments appearing within an arbitrary argument framework induced from the probabilistic framework.

Weights can be also interpreted as subjective beliefs [20]. For example, a weight of $w \in (0, 1]$ on the attack of argument a_1 on argument a_2 might be understood as the belief that (a decision-maker considers) a_2 is false when a_1 is true. This belief could be modelled using probability [20] as well.

The *Fuzzy Argumentation* approach presented in [29] enriches the expressive power of the classical argumentation model by allowing to represent the relative strength of the attack relations between arguments, as well as the degree to which arguments are accepted.

In addition, the weighted semiring $\langle \mathbb{R}^+ \cup \{\infty\}, \min, +, \infty, 0 \rangle$ can model a generic "cost" for the attacks: for example, the number of votes in support of the attack [20], which consequently needs to be minimised.

5 Conclusions and Future Work

In this work we have defined a new notion of defence for WAAFs. Since defence is collective in the literature (i.e., it considers all the counter-attacks from $\mathcal{B}$ as a whole), our main motivation is to provide a similar view also for all the attacks from a to $\mathcal{B}$, here considered by summing all the attacks weights together. In addition, by casting similar proposals [18, 26] in the same parametric algebraic-framework, it is possible to show all their relations in detail.

Finally, we will study how the presented framework can be used to model *ranking-based* semantics [15], where a (partial) preference order is defined among the arguments of an AAF. A first step in this direction has been already moved in [8, 9], but our plan is to generalise different ranking-based functions by using our semiring-based framework with weights computed as *Shapely values*.

Acknowledgements We would like to thank Daniele Pirolandi, Fabio Rossi, Francesco Santini and Carlo Taticchi for unvaluable discussions and commnents that lead to the results summarised in this short paper. This research is partially supported by project "Argumentation 360" and "RACRA18" (Funded by Dept. of Mathematics and Computer Science, University of Perugia).

References

1. Amgoud, L., Ben-Naim, J., Doder, D., Vesic, S.: Ranking arguments with compensation-based semantics. In: Principles of Knowledge Representation and Reasoning: Proceedings of the Fifteenth International Conference, KR, pp. 12–21. AAAI Press (2016)
2. Amgoud, L., Cayrol, C.: On the acceptability of arguments in preference-based argumentation. In: UAI '98: Proceedings of the Fourteenth Conference on Uncertainty in Artificial Intelligence, pp. 1–7. Morgan Kaufmann (1998)
3. Bench-Capon, T.J.M.: Persuasion in practical argument using value-based argumentation frameworks. J. Log. Comput. **13**(3), 429–448 (2003)
4. Benedetti, Irene, Bistarelli, S.: From argumentation frameworks to voting systems and back. Fundamenta Informaticae **150**(1), 25–48 (2017)
5. Bistarelli, S., Montanari, U., Rossi, F.: Semiring-based constraint solving and optimization. J. ACM **44**(2), 201–236 (1997)
6. Bistarelli, S., Santini, F.: Conarg: a constraint-based computational framework for argumentation systems. In: IEEE 23rd International Conference on Tools with Artificial Intelligence, ICTAI 2011, pp. 605–612. IEEE (2011)
7. Bistarelli, S.: Semirings for Soft Constraint Solving and Programming. Lecture Notes in Computer Science, vol. 2962. Springer (2004)
8. Bistarelli, S., Faloci, F., Santini, F., Taticchi, C.: A tool for ranking arguments through voting-games power indexes. In: Alberto, C., Eugenio, G.O. (eds.) Proceedings of the 34th Italian Conference on Computational Logic, Trieste, Italy, June 19-21, 2019., vol. 2396 of CEUR Workshop Proceedings, pp. 193–201. CEUR-WS.org (2019)
9. Bistarelli, S., Giuliodori, P., Santini, F., Taticchi, C.: A cooperative-game approach to share acceptability and rank arguments. In: Dondio, P., Longo, L. (eds.) Proceedings of the 2nd Workshop on Advances In Argumentation In Artificial Intelligence, co-located with XVII International Conference of the Italian Association for Artificial Intelligence, AI^3@AI*IA 2018, 20-23 November 2018, Trento, Italy, vol. 2296 of CEUR Workshop Proceedings, pp. 86–90. CEUR-WS.org (2018)

10. Bistarelli, S., Rossi, F., Francesco, S.: A collective defence against grouped attacks for weighted abstract argumentation frameworks. In: Proceedings of the Twenty-Ninth International Florida Artificial Intelligence Research Society Conference, FLAIRS, pp. 638–643. AAAI Press (2016)
11. Bistarelli, S., Rossi, F., Santini, F.: A relaxation of internal conflict and defence in weighted argumentation frameworks. In: Logics in Artificial Intelligence—15th European Conference, JELIA, vol. 10021 of Lecture Notes in Computer Science, pp. 127–143. Springer (2016)
12. Bistarelli, S., Rossi, F., Francesco, S.: A novel weighted defence and its relaxation in abstract argumentation. Int. J. Approx. Reasoning **92**, 66–86 (2018)
13. Bistarelli, S., Santini, F.: A common computational framework for semiring-based argumentation systems. In: Coelho, H., Studer, R., Wooldridge, M. (eds.) ECAI, vol. 215 of Frontiers in Artificial Intelligence and Applications, pp. 131–136. IOS Press (2010)
14. Bistarelli, S., Santini, F.: Well-foundedness in weighted argumentation frameworks. In: Calimeri, F., Leone, N., Manna, M. (eds.) Logics in Artificial Intelligence—16th European Conference, JELIA 2019, Rende, Italy, May 7-11, 2019, Proceedings, vol. 11468 of Lecture Notes in Computer Science, pp. 69–84. Springer (2019)
15. Bonzon, E., Delobelle, J., Konieczny, S., Maudet, N.: A comparative study of ranking-based semantics for abstract argumentation. In: Proceedings of the Thirtieth AAAI Conference on Artificial Intelligence, pp. 914–920. AAAI Press (2016)
16. Cayrol, C., Devred, C., Lagasquie-Schiex, M.C.: Acceptability semantics accounting for strength of attacks in argumentation. In: ECAI 2010—19th European Conference on Artificial Intelligence, vol. 215, pp. 995–996. IOS Press (2010)
17. Cayrol, C., Lagasquie-Schiex, M.C.: From preferences over arguments to preferences over attacks in abstract argumentation: a comparative study. In: 2013 IEEE 25th International Conference on Tools with Artificial Intelligence, pp. 588–595. IEEE Computer Society (2013)
18. Coste-Marquis, S., Konieczny, S., Marquis, P., Akli Ouali, M.: Weighted attacks in argumentation frameworks. In: Principles of Knowledge Representation and Reasoning: Proceedings of the Thirteenth International Conference, KR, pp. 593–597. AAAI Press (2012)
19. Dung, P.M.: On the acceptability of arguments and its fundamental role in nonmonotonic reasoning, logic programming and n-person games. Artif. Intell. **77**(2), 321–357 (1995)
20. Dunne, P.E., Hunter, A., McBurney, P., Parsons, S., Wooldridge, M.: Inconsistency tolerance in weighted argument systems. In: Conference on Autonomous Agents and Multiagent Systems, pp. 851–858. IFAAMS (2009)
21. Egilmez, S., Martins, J., Leite, J.: Extending social abstract argumentation with votes on attacks. In: Theory and Applications of Formal Argumentation—Second International Workshop, TAFA, vol. 8306, pp. 16–31. Springer (2013)
22. Grossi, D., Modgil, S.: On the graded acceptability of arguments. In: Proceedings of the Twenty-Fourth International Joint Conference on Artificial Intelligence, IJCAI, pp. 868–874. AAAI Press (2015)
23. Kaci, S., Labreuche, C.: Arguing with valued preference relations. In: Symbolic and Quantitative Approaches to Reasoning with Uncertainty—11th European Conference, ECSQARU, vol. 6717 of Lecture Notes in Computer Science, pp. 62–73. Springer (2011)
24. Leite, J., Martins, J.: Social abstract argumentation. In: IJCAI 2011, Proceedings of the 22nd International Joint Conference on Artificial Intelligence, pp. 2287–2292. IJCAI/AAAI (2011)
25. Li, H., Oren, N., Norman, T.J.: Probabilistic argumentation frameworks. In: Theorie and Applications of Formal Argumentation—First International Workshop, TAFA, vol. 7132 of Lecture Notes in Computer Science, pp. 1–16. Springer (2012)
26. Martínez, D.C., García, A.J., Simari, G.R.: An abstract argumentation framework with varied-strength attacks. In: Principles of Knowledge Representation and Reasoning: Proceedings of the Eleventh International Conference, pp. 135–144. AAAI Press (2008)
27. Modgil, S.: Reasoning about preferences in argumentation frameworks. Artif. Intell. **173**(9–10), 901–934 (2009)
28. Nielsen, S.H., Parsons, S.: A generalization of dung's abstract framework for argumentation: arguing with sets of attacking arguments. In: Maudet, N., Parsons, S., Rahwan, I. (eds.) Argumentation in Multi-Agent Systems, Third International Workshop, ArgMAS 2006, Hakodate,

Japan, May 8, 2006, Revised Selected and Invited Papers, vol. 4766 of Lecture Notes in Computer Science, pp. 54–73. Springer (2006)

29. Schroeder, M., Schweimeier, R.: Fuzzy argumentation for negotiating agents. In: The First International Joint Conference on Autonomous Agents & Multiagent Systems, AAMAS, pp. 942–943. ACM (2002)

Further Reading

1. Bistarelli, S., Pirolandi, D., Santini, F.: Solving weighted argumentation frameworks with soft constraints. In: Larrosa, J., O'Sullivan, B. (eds.) Recent Advances in Constraints—14th Annual ERCIM International Workshop on Constraint Solving and Constraint Logic Programming, CSCLP 2009, Barcelona, Spain, June 15–17, 2009, Revised Selected Papers, vol. 6384 of Lecture Notes in Computer Science, pp. 1–18. Springer (2009)

2. Bistarelli, S., Pirolandi, D., Santini, F.: Solving weighted argumentation frameworks with soft constraints. In: Faber, W., Leone, N. (eds.) Proceedings of the 25th Italian Conference on Computational Logic, Rende, Italy, July 7–9, 2010, vol. 598 of CEUR Workshop Proceedings. CEUR-WS.org (2010)

3. Bistarelli, S., Rossi, F., Santini, F.: Conarg: a tool for classical and weighted argumentation. In: Baroni, P., Gordon, T.F., Scheffler, T., Stede, M. (eds.) Computational Models of Argument—Proceedings of COMMA 2016, Potsdam, Germany, 12–16 September, 2016, vol. 287 of Frontiers in Artificial Intelligence and Applications, pp. 463–464. IOS Press (2016)

4. Bistarelli, S., Santini, F.: A Hasse diagram for weighted sceptical semantics with a unique-status grounded semantics. In: Logic Programming and Nonmonotonic Reasoning—14th International Conference, LPNMR, vol. 10377 of Lecture Notes in Computer Science, pp. 49–56. Springer (2017)

5. Bistarelli, S., Santini, F.: Some thoughts on well-foundedness in weighted abstract argumentation. In: Thielscher, M., Toni, F., Wolter, F. (eds.) Principles of Knowledge Representation and Reasoning: Proceedings of the Sixteenth International Conference, KR 2018, Tempe, Arizona, 30 October–2 November 2018, pp. 623–624. AAAI Press (2018)

6. Bistarelli, S., Tappini, A., Taticchi, C.: A matrix approach for weighted argumentation frameworks. In: Brawner, K., Rus, V. (eds.) Proceedings of the Thirty-First International Florida Artificial Intelligence Research Society Conference, FLAIRS 2018, Melbourne, Florida, USA. May 21–23 2018, pp. 507–512. AAAI Press (2018)

Modeling and Specification of Nondeterministic Fuzzy Discrete-Event Systems

Yongzhi Cao, Yoshinori Ezawa, Guoqing Chen and Haiyu Pan

Abstract Most of the published research on fuzzy discrete-event systems (FDESs) has focused on systems that are modeled as deterministic fuzzy automata. In fact, nondeterminism in FDESs occurs in many practical situations and can be used to represent underspecification or incomplete information. In this paper, we pay attention to the modeling and specification of nondeterministic FDESs (NFDESs). We model NFDESs by a new kind of fuzzy automata. To describe adequately the behavior of NFDESs, we introduce the concept of bisimulation, which is a finer behavioral measure than fuzzy language equivalence. Further, we propose the notion of nondeterministic fuzzy specifications (NFSs) to specify the behavior of NFDESs and introduce a satisfaction relation between NFDESs and NFSs. If such a relation exists, then at least one knows that there is no unwanted behavior in the system.

1 Introduction

A discrete-event system (DES) is a dynamic system whose state space is discrete and whose states can only change as a result of events occurring instantaneously

Y. Cao (✉)
Key Laboratory of High Confidence Software Technologies (MOE),
Department of Computer Science and Technology, Peking University,
Beijing 100871, China
e-mail: caoyz@pku.edu.cn

Y. Ezawa
Faculty of Informatics, Kansai University, Osaka 569-1095, Japan
e-mail: ezawa@res.kutc.kansai-u.ac.jp

G. Chen
School of Economics and Management, Tsinghua University, Beijing 100084, China
e-mail: chengq@sem.tsinghua.edu.cn

H. Pan
Guangxi Key Laboratory of Trusted Software,
Guilin University of Electronic Technology, Guilin 541004, China
e-mail: phyu76@126.com

© Springer Nature Switzerland AG 2020
M. Ceberio and V. Kreinovich (eds.), *Decision Making under Constraints*,
Studies in Systems, Decision and Control 276,
https://doi.org/10.1007/978-3-030-40814-5_6

45

over time. In the framework of Ramadge and Wonham [34], a DES is modeled as a deterministic or nondeterministic automaton that executes state transitions in response to a sequence of events occurring asynchronously. A common feature of these models is that their state transitions are crisp once that corresponding events occur. Nevertheless, there are situations in which the state transitions of some systems [21, 35, 36] are always somewhat uncertain.

The consideration above has called for a finer model where uncertainty is formally described. The approaches to this model can be naturally divided into two groups. One is concerned with probabilistic uncertainty and in the resulting model of probabilistic automata, both nondeterministic and probabilistic choices are present at each transition [12, 17, 19, 20, 27]. The other deals with the modeling of DESs with fuzzy uncertainty by using fuzzy automata, which was initiated by Lin and Ying [21], and the resultant systems are called fuzzy discrete-event systems (FDESs). In FDESs, every classical transition is replaced by a fuzzy transition whose target is a possibility distribution over states. So far, a large number of studies such as event-based control [5, 11, 24, 30], state-based control [7, 10, 22], and observability and decentralized control [6, 31] have been conducted to explore the theory of FDESs. At the same time, many interesting applications to biomedicine [35, 36], mobile robotics [15, 25], failure prognosis [1], and decision making [23] have been reported. The present work falls into the second group.

It is worth noting that the models of FDESs proposed in the literature are extensions of DESs to deterministic uncertainty, but have no nondeterminism, in the sense that the choice of (fuzzy) event determines the next transition. As is well known, nondeterminism is essential for modeling scheduling freedom, implementation freedom, the external environment, and incomplete information (see, for example, [14]). In the literature, nondeterminism has already been taken into account in both classical DESs [13, 16, 37, 38] and probabilistic DESs [12]. In practice, not all DESs with fuzzy uncertainty are deterministic either. For example, in [6, 35, 36] FDESs are used to model the potency of a regimen for patients, where the potency is typically described like the following: The possibility of transferring from "poor" (P) state to "fair" (F) state is 0.9 and to "excellent" (E) state is 0.2 after a regimen. On the other hand, it is a common knowledge that applying the same treatment planning to different patients, even the identical patient at various times, with the same health status may give rise to diverse effects, say, it is possible to change a patient into F state with possibility degree 0.2 while keeping P state with possibility degree 0.9 after the regimen. In other words, the same regimen could result in different effects. As a result, nondeterminism appears. Although each individual effect is represented by a fuzzy set, note that neither probabilities nor possibility degrees among the effects are assigned, either because the information available is not enough to produce a meaningful estimate of the quantitative values or because the quantitative information is not really valuable for a specific patient. Clearly, neither probabilistic DESs nor FDESs can model such situations.

The above observation motivates us to introduce a formalism for the modeling and specification of FDESs with nondeterminism, which we will refer to as nondeterministic FDESs (NFDESs), in this paper. We model NFDESs by nondeterministic fuzzy automata, which are a new kind of fuzzy automata introduced by Cao and Ezawa in [3]. Unlike the conventional FDESs, fuzzy languages do not turn out to be adequate for describing the behavior of NFDESs due to the nondeterminism. We thus appeal to a finer behavioral measure—bisimulation proposed by Park [28] and Milner [26]. Two NFDESs are bisimilar if their initial states are related by a certain bisimulation. Intuitively, bisimilar NFDESs behave in the same way in the sense that they match each other's moves. We examine some properties of bisimulation. In particular, we show that bisimilarity is preserved by the product and parallel composition operators. Further, to specify the behavior of NFDESs, we propose the notion of nondeterministic fuzzy specifications (NFSs) and introduce a satisfaction relation between NFDESs and NFSs which requires that each move of the system can be simulated by the corresponding specification.

The remainder of the paper is structured as follows. After reviewing some basics of the conventional FDESs in Sect. 2, we embark on the model of NFDESs in Sect. 3 and address the notion of bisimulation and related properties in Sect. 4. Section 5 is devoted to NFSs and the satisfaction relation. We conclude the paper in Sect. 6 with a brief discussion on the future research.

2 Fuzzy Discrete-Event Systems

In this section, we briefly review the formulation of FDESs.

Let X be a universal set. Recall that a fuzzy set A, or rather a *fuzzy subset A of X*, is defined by a function assigning to each element x of X a value $A(x)$ in $[0, 1]$. Such a function is called a *membership function*, and any fuzzy set of X can be regarded as a possibility distribution on X. When there are only finitely many elements, say $x_1, x_2, \ldots, x_n$, having positive values, we may write A in Zadeh's notation as

$$A = \frac{A(x_1)}{x_1} + \frac{A(x_2)}{x_2} + \cdots + \frac{A(x_n)}{x_n}.$$

With this notation, $1/x$ is a singleton in X. A fuzzy set is said to be *empty*, denoted $\emptyset$, if its membership function is identically zero on X. We write $\mathscr{F}(X)$ for the set of all fuzzy sets of X and $\mathscr{P}(X)$ for the power set of X.

For any family λ_i, $i \in I$, of elements of $[0, 1]$, we write $\vee_{i \in I} \lambda_i$ or $\vee \{\lambda_i \mid i \in I\}$ for the supremum of $\{\lambda_i \mid i \in I\}$, and $\wedge_{i \in I} \lambda_i$ or $\wedge \{\lambda_i \mid i \in I\}$ for the infimum. For any given $\lambda \in [0, 1]$ and $\mu \in \mathscr{F}(X)$, the fuzzy set $\lambda \cdot \mu$ of X is defined by $(\lambda \cdot \mu)(x) = \lambda \wedge \mu(x)$ for any $x \in X$. Given $\mu \in \mathscr{F}(X)$ and $\eta \in \mathscr{F}(Y)$, we write $\mu \wedge \eta$ for the fuzzy set of $X \times Y$ that is defined by $(\mu \wedge \eta)(x, y) = \mu(x) \wedge \eta(y)$ for all $(x, y) \in X \times Y$. For any $\mu \in \mathscr{F}(X)$ and $U \subseteq X$, the notation $\mu(U)$ stands for $\vee_{x \in U} \mu(x)$.

In the framework of Ramadge and Wonham [34], a DES is usually represented by a deterministic or nondeterministic finite automaton. To capture the uncertainty appearing in states and state transitions of DESs, two kinds of fuzzy automata known as max-min and max-product automata have been adopted [5, 21, 30]. In the paper, we restrict ourselves to the following max-min automaton model from [5].

Definition 1 A fuzzy discrete-event system (FDES) is modeled by a max-min automaton $G = (Q, \Sigma, \delta, q_0)$, where

(1) Q is a finite or infinite set of states;
(2) Σ is a finite set of events;
(3) $\delta : Q \times \Sigma \longrightarrow \mathscr{F}(Q)$ is a function, called a fuzzy transition function;
(4) $q_0 \in Q$ is the initial state.

Intuitively, for any $p, q \in Q$ and $a \in \Sigma$, we can interpret $\delta(p, a)(q)$ as the possibility degree to which the FDES G in state p and with the occurrence of event a may enter state q. The behavior of G is usually described by the *fuzzy language* $\mathscr{L}(G)$ generated by G. It is a fuzzy set of Σ^*, the set of all finite strings constructed by concatenation of elements of Σ, and defined as $\mathscr{L}(G)(s) = \vee_{q \in Q} \delta(q_0, s)(q)$, where $\delta(q_0, s)$ is defined inductively as follows:

$$\delta(q_0, \varepsilon)(q) = \begin{cases} 1, & \text{if } q = q_0 \\ 0, & \text{otherwise} \end{cases}$$

$$\delta(q_0, sa)(q) = \vee_{p \in Q}[\delta(q_0, s)(p) \wedge \delta(p, a)(q)]$$

for all $s \in \Sigma^*$ and $a \in \Sigma$.

3   Nondeterministic Fuzzy Discrete-Event Systems

In this section, we first introduce the formal model of NFDESs and then present the product and parallel composition operators for composing NFDESs. Finally, for comparison, we introduce the notion of fuzzy languages generated by NFDESs.

We model FDESs with nondeterminism as follows.

Definition 2 A nondeterministic fuzzy discrete-event system (NFDES) is modeled by a nondeterministic fuzzy automaton $G = (Q, \Sigma, \delta, q_0)$, where

(1) Q is a finite or infinite set of states;
(2) Σ is a finite set of events;
(3) $\delta : Q \times \Sigma \longrightarrow \mathscr{P}(\mathscr{F}(Q))$ is a fuzzy transition function which gives rise to a set $\delta(q, a)$ of possibility distributions on Q, for each $q \in Q$ and $a \in \Sigma$;
(4) $q_0 \in Q$ is the initial state.

We use $q \xrightarrow{a} \mu$ to denote that $\mu \in \delta(q, a)$, and refer to $q \xrightarrow{a} \mu$ as a *transition*. If $\mu = \emptyset$, it is usually interpreted that no such transition exists. For each $q \in Q$, we write Σ_q for $\{a \in \Sigma \mid \exists \mu \neq \emptyset, q \xrightarrow{a} \mu\}$, where the transitions represent the nondeterministic alternatives in the state q. The nondeterministic choices are beyond

Fig. 1 Two NFDESs
generating the same fuzzy
language

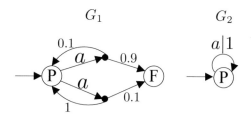

the control of the NFDES and are supposed to be resolved by the environment, whereas the possibility degrees of entering next states are made by the system itself, according to the underlying possibility distribution. More concretely, if $\delta(q, a) = \{\mu_i \mid i \in I\}$, then the NFDES in state q and with the occurrence of event a yields some possibility distribution in $\{\mu_i \mid i \in I\}$. Which one is chosen depends on the environment. If μ_j is chosen, then the NFDES may enter each state q' in Q with possibility degree $\mu_j(q')$.

Clearly, NFDESs have FDESs as a special case. In contrast with NFDESs, the FDES model does not allow nondeterministic choices between transitions involving the same event. It should be pointed out that the nondeterministic fuzzy automaton in the above definition is somewhat different from that in [3]. We have allowed the state set Q to be infinite because sometimes Q consists of all fuzzy states on some underlying state set which is not finite. In addition, the fuzzy set of final states has not been included since in the work we are not concerned with the properties relevant to final states.

Different semantics or notions of behavior can be given to NFDESs. Among others, let us first consider fuzzy languages, since they have been extensively used to describe the behavior of FDESs [5, 6, 11, 24, 30]. The next definition follows from [3, Definition 4] by considering all states as finial states.

Definition 3 Let $G = (Q, \Sigma, \delta, q_0)$ be an NFDES. The fuzzy language $\mathscr{L}(G)$ generated by G is a fuzzy set of Σ^* with the membership function defined by $\mathscr{L}(G)(s) = \vee_{\mu \in \delta(q_0, s)} \vee_{q \in Q} \mu(q)$ for all $s \in \Sigma^*$, where $\delta(q_0, s)$ is defined inductively as follows:

$$\delta(q_0, \varepsilon) = \{1/q_0\}$$
$$\delta(q_0, sa) = \{\mu(p) \cdot \mu' \mid \mu \in \delta(q_0, s), p \in Q, \mu' \in \delta(p, a)\}$$

for all $s \in \Sigma^*$ and $a \in \Sigma$.

Following the notion of fuzzy languages for FDESs, max-min operator has been adopted in the above definition. There is no difficulty to check that this definition is consistent with that of fuzzy languages for FDESs in the sense that viewing an FDES G as an NFDES and using the above definition to compute $\mathscr{L}(G)$ gives the same result as directly using fuzzy language definition of FDESs.

We may represent an NFDES via its transition diagram. Like FDESs, the states are in circles. Now, each transition is depicted via two parts: an arrow for nondeterministic choice and a bunch of arrows for the possibility degrees of entering next states. For example, $\delta(\text{P}, a) = \{0.1/\text{P} + 0.9/\text{F}, 1/\text{P} + 0.1/\text{F}\}$ is depicted as shown on the left of Fig. 1. In particular, if $q \xrightarrow{a} \gamma/p$, then we often simply depict it like the right of Fig. 1.

For the NFDESs in Fig. 1, a routine computation shows that $\mathscr{L}(G_1)(s) = 1 = \mathscr{L}(G_2)(s)$ for all $s \in \{a\}^*$. Consequently, G_1 and G_2 are equivalent if the behavior of NFDESs is characterized by fuzzy languages. However, we can sense that the two systems are distinctly different. For example, G_1 may reach the state F with possibility degree 0.9, while G_2 is always in the state P. More seriously, fuzzy language equivalent systems can have different deadlocking behavior (inability to proceed). These observations suggest that using fuzzy language equivalence to compare the behavior of NFDESs is too coarse sometimes, which leads us to consider bisimulation, a finer behavioral measure, in the subsequent section.

Like classical DESs, one can build an overall NFDES by building its component systems first and then composing them by some operators. As an example, we introduce a product operator and a parallel composition operator. To this end, it is convenient to have one more notation. Without loss of generality, we consider NFDESs with the same set of events.

Let $G_i = (Q_i, \Sigma, \delta_i, q_{0i})$, $i = 1, 2$, be an NFDES. The *product* of G_1 and G_2 is $G_1 \times G_2 = (Q_1 \times Q_2, \Sigma, \delta, (q_{01}, q_{02}))$, where $\delta : (Q_1 \times Q_2) \times \Sigma \longrightarrow \mathscr{P}$ $(\mathscr{F}(Q_1 \times Q_2))$ is given by

$$\delta((p, q), a) = \begin{cases} \{\mu \wedge \eta \mid \mu \in \delta_1(p, a), \eta \in \delta_2(q, a)\}, & \text{if } a \in \Sigma_p \cap \Sigma_q \\ \emptyset, & \text{otherwise.} \end{cases}$$

The product requires that the components are strictly synchronous, while the following parallel composition operator is asynchronous in the sense that the components can either synchronize or act independently. Formally, the *parallel composition* of G_1 and G_2 is $G_1|G_2 = (Q_1 \times Q_2, \Sigma, \delta, (q_{01}, q_{02}))$, where for all $(p, q) \in Q_1 \times Q_2$ and $a \in \Sigma$,

$$\delta((p, q), a) = \begin{cases} \{\mu \wedge \eta \mid \mu \in \delta_1(p, a), \eta \in \delta_2(q, a)\}, & \text{if } a \in \Sigma_p \cap \Sigma_q \\ \{\mu \wedge (1/q) \mid \mu \in \delta_1(p, a)\}, & \text{if } a \in \Sigma_p \backslash \Sigma_q \\ \{(1/p) \wedge \eta \mid \eta \in \delta_2(q, a)\}, & \text{if } a \in \Sigma_q \backslash \Sigma_p \\ \emptyset, & \text{otherwise.} \end{cases}$$

Clearly, both $G_1 \times G_2$ and $G_1|G_2$ are again NFDESs.

4 Bisimulation

Bisimulation is a well-known equivalence relation for process algebra [26, 28] and has been extended to probabilistic [18] and some fuzzy systems [2, 9, 29, 32]. In this section, we introduce the notion of bisimulation for NFDESs. For generality, we first recall bisimulation for NFDESs without initial states which are referred to as (nondeterministic) *fuzzy-transition systems* (FTSs) [4, 8] and then apply it to NFDESs.

Definition 4 Let (Q, Σ, δ) be an FTS. An equivalence relation R on Q is called a bisimulation if for any $(p, q) \in R$ and $a \in \Sigma$, $p \xrightarrow{a} \mu$ implies $q \xrightarrow{a} \eta$ for some η such that $\mu(C) = \eta(C)$ for every equivalence class $C \in Q/R$. Two states p and q are bisimilar, written $p \sim q$, if $(p, q) \in R$ for some bisimulation R.

The idea behind the bisimulation is as follows. As bisimilar states are considered the same, it does not matter which state within a bisimulation class is reached. Therefore, a bisimulation relation should compare the possibility degree to reach an equivalence class and not the possibility degree to reach a single state.

To state the next result, we need one more notation. The *transitive closure* of a binary relation R on Q is the minimal transitive relation R^* on Q that contains R. Thus, if R_1 and R_2 are two equivalence relations on Q, then so is $(R_1 \cup R_2)^*$.

Proposition 1 *Let $\sim = (\cup_i R_i)^*$, where R_i is a bisimulation on Q. Then $\sim$ is the largest bisimulation on Q.*

Proof It suffices to show that $\sim$ is a bisimulation on Q. Suppose that $(p, q) \in \sim$ and $p \xrightarrow{a} \mu_0$. Then there are $p_0, \ldots, p_m \in Q$ and bisimulations $R_{1'}, \ldots, R_{m'}$ such that $p = p_0$, $q = p_m$, and $(p_{i-1}, p_i) \in R_{i'}$ for $i = 1, \ldots, m$. By definition, there exist $\mu_1, \ldots, \mu_m$ such that $p_i \xrightarrow{a} \mu_i$ and $\mu_{i-1}(C_{i'}) = \mu_i(C_{i'})$ for $i = 1, \ldots, m$ and any $C_{i'} \in Q/R_{i'}$. Note that for any $C \in Q/\sim$ and $R_{i'}$, it follows from $R_{i'} \subseteq \sim$ that $C = \cup_j C_{i'j}$ for some $C_{i'j} \in Q/R_{i'}$, and moreover, $\cup_j C_{i'j}$ is a disjoint union. We thus have that $\mu_{i-1}(C) = \mu_{i-1}(\cup_j C_{i'j}) = \sum_j \mu_{i-1}(C_{i'j}) = \sum_j \mu_i(C_{i'j}) = \mu_i(C)$ for each $i = 1, \ldots, m$. This means that $\mu_0(C) = \mu_m(C)$. Hence, $\sim$ is a bisimulation. ∎

The largest bisimulation $\sim$ is called *bisimilarity*. We now consider the bisimulations on NFDESs.

Definition 5 Let $G_i = (Q_i, \Sigma, \delta_i, q_{0i})$, $i = 1, 2$, be an NFDES. We say that G_1 and G_2 are bisimilar, denoted $G_1 \sim G_2$, if $q_{01} \sim q_{02}$ in the FTS $(Q_1 \dot\cup Q_2, \Sigma, \delta)$, where $Q_1 \dot\cup Q_2$ is the disjoint union of Q_1 and Q_2, and $\delta : (Q_1 \dot\cup Q_2) \times \Sigma \longrightarrow \mathscr{P}(\mathscr{F}(Q_1 \dot\cup Q_2))$ is defined by

$$\delta(q, a) = \begin{cases} \delta_1(q, a), & \text{if } q \in Q_1 \\ \delta_2(q, a), & \text{if } q \in Q_2. \end{cases}$$

Fig. 2 Three NFDESs:
$G_1 \sim G_2 \not\sim G_3$.

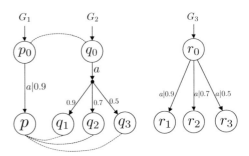

Two bisimilar NFDESs behave in the same way in the sense that one system simulates the other and vice-versa. It should be remarked that in the above definition, we have viewed each possibility distribution on $Q_i, i = 1, 2$, as a possibility distribution on $Q_1 \dot{\cup} Q_2$ in the obvious way.

Example 1 Observe that G_1 and G_2 in Fig. 2 are bisimilar because the smallest equivalence relation containing $\{(p_0, q_0), (p, q_1), (p, q_2), (p, q_3)\}$ is a bisimulation. However, G_2 and G_3 are not bisimilar because there is no bisimulation that contains (q_0, r_0). In fact, the transitions $r_0 \xrightarrow{a} 0.7/r_2$ and $r_0 \xrightarrow{a} 0.5/r_3$ in G_3 cannot be matched by the unique transition $q_0 \xrightarrow{a} 0.9/q_1 + 0.7/q_2 + 0.5/q_3$ in G_2. By the same token, G_1 and G_3 are not bisimilar either.

Bisimulation has a number of pleasing properties. For example, bisimilar NFDESs generate the same fuzzy language.

Proposition 2 *Let* $G_i = (Q_i, \Sigma, \delta_i, q_{0i})$, $i = 1, 2$, *be an NFDES. If* $G_1 \sim G_2$, *then* $\mathscr{L}(G_1) = \mathscr{L}(G_2)$.

Proof It is sufficient to show that $\mathscr{L}(G_1)(s) = \mathscr{L}(G_2)(s)$ for any string $s \in \Sigma^*$, which can be easily verified by induction on the length of s. We thus omit the details here.

The following observation shows that the product and parallel composition are commutative with respect to $\sim$.

Proposition 3 *Let* $G_i = (Q_i, \Sigma, \delta_i, q_{0i})$, $i = 1, 2$, *be an NFDES. Then we have the following.*

(1) $G_1 \times G_2 \sim G_2 \times G_1$.
(2) $G_1|G_2 \sim G_2|G_1$.

Proof Let R be the smallest equivalence relation containing $\{((p, q), (q, p)) \mid (p, q) \in Q_1 \times Q_2\}$. It is straightforward to verify that R is a bisimulation in both cases. We omit the details.

The following theorem shows that bisimilarity is preserved by the product and parallel composition operators. As a result, we see by the theorem that $G_1 \not\sim G_2$ if we can find some G_3 such that either $G_1 \times G_3 \not\sim G_2 \times G_3$ or $G_1|G_3 \not\sim G_2|G_3$.

Theorem 1 *Let $G_i = (Q_i, \Sigma, \delta_i, q_{0i})$, $i = 1, 2, 3$, be an NFDES. If $G_1 \sim G_2$, then we have the following:*

(1) $G_1 \times G_3 \sim G_2 \times G_3$.
(2) $G_1|G_3 \sim G_2|G_3$.

Proof We only prove (2); one can prove (1) by imitating Case 1 below. Let us write R' for the bisimilarity on $(Q_1 \dot{\cup} Q_2, \Sigma, \delta)$. Consider $R = \{((p, r), (q, r)) \mid (p, q) \in R', r \in Q_3\}$. Clearly, R is an equivalence relation on $(Q_1 \times Q_3) \dot{\cup} (Q_2 \times Q_3)$, and moreover, $((q_{01}, q_{03}), (q_{02}, q_{03})) \in R$. To show that $G_1|G_3 \sim G_2|G_3$, it remains to verify that R is a bisimulation. For any $((p, r), (q, r)) \in R$ and $a \in \Sigma$, if $(p, r) \xrightarrow{a} \mu$, it suffices to show that there is $(q, r) \xrightarrow{a} \eta$ such that $\mu(C) = \eta(C)$ for any $C \in ((Q_1 \times Q_3) \dot{\cup} (Q_2 \times Q_3))/R$. Note that such an equivalence class C gives exactly an equivalence class $C' = \{p \mid (p, r) \in C\} \in (Q_1 \dot{\cup} Q_2)/R'$ and a subset $C'' = \{r \mid (p, r) \in C\}$ of Q_3. By the definition of parallel composition, four cases need to be considered.

   Case 1: $a \in \Sigma_p \cap \Sigma_r$. In this case, there are $p \xrightarrow{a} \mu_p$ and $r \xrightarrow{a} \theta_r$ with $\mu_p \wedge \theta_r = \mu$. Since $(p, q) \in R'$, there exists $q \xrightarrow{a} \eta_q$ such that $\mu_p(C) = \eta_q(C)$ for any $C \in (Q_1 \dot{\cup} Q_2)/R'$. Taking $\eta = \eta_q \wedge \theta_r$, it yields a transition $(q, r) \xrightarrow{a} \eta$. Moreover, for any $C \in ((Q_1 \times Q_3) \dot{\cup} (Q_2 \times Q_3))/R$, we have that $\mu(C) = (\mu_p \wedge \theta_r)(C) = \mu_p(C') \wedge \theta_r(C'') = \eta_q(C') \wedge \theta_r(C'') = (\eta_q \wedge \theta_r)(C) = \eta(C)$, namely, $\mu(C) = \eta(C)$, as desired.

   Case 2: $a \in \Sigma_p \backslash \Sigma_r$. In this case, there is only one transition $p \xrightarrow{a} \mu_p$, which forces that $\mu = \mu_p \wedge (1/r)$. As $(p, q) \in R'$, there exists $q \xrightarrow{a} \eta_q$ such that $\mu_p(C) = \eta_q(C)$ for any $C \in (Q_1 \dot{\cup} Q_2)/R'$. Taking $\eta = \eta_q \wedge (1/r)$, it yields a transition $(q, r) \xrightarrow{a} \eta$. Moreover, for any $C \in ((Q_1 \times Q_3) \dot{\cup} (Q_2 \times Q_3))/R$, we have that $\mu(C) = (\mu_p \wedge (1/r))(C) = \mu_p(C') \wedge (1/r)(C'') = \eta_q(C') \wedge (1/r)(C'') = (\eta_q \wedge (1/r))(C) = \eta(C)$, i.e., $\mu(C) = \eta(C)$.

   Case 3: $a \in \Sigma_r \backslash \Sigma_p$. This case is symmetric to Case 2.
   Case 4: $a \notin \Sigma_p \cup \Sigma_r$. There is nothing to prove.

5 Nondeterministic Fuzzy Specification

In this section, we generalize the definition of NFDESs to nondeterministic fuzzy specifications (NFSs) and introduce the notion of satisfaction relation between NFDESs and NFSs. If there is a satisfaction relation between an actual system and its specification, then at least one knows that there is no unwanted behavior in the actual system. In that case, the system can be viewed as an implementation of the specification as each move of the system is allowed by the specification.

   Prior to the definition of NFSs, we introduce a notation. We use $\widehat{\mathscr{F}}(Q)$ to denote the set of mappings from Q to the power set of $[0, 1]$, namely, $\widehat{\mathscr{F}}(Q) = \{\widehat{\mu} \mid \widehat{\mu} : Q \longrightarrow \mathscr{P}([0, 1])\}$.

Definition 6 A nondeterministic fuzzy specification (NFS) is a four-tuple $S = (Q, \Sigma, \widehat{\delta}, q_0)$, where

(1) Q is a finite or infinite set of states;
(2) Σ is a finite set of events;
(3) $\widehat{\delta} : Q \times \Sigma \longrightarrow \mathscr{P}(\widehat{\mathscr{F}}(Q))$ is a set-valued transition function which gives a set $\widehat{\delta}(q, a)$ of set-valued possibility distributions on Q, for each $q \in Q$ and $a \in \Sigma$;
(4) $q_0 \in Q$ is the initial state.

Analogous to NFDESs, we use $q \overset{a}{\longrightarrow} \widehat{\mu}$ to denote that $\widehat{\mu} \in \widehat{\delta}(q, a)$, and refer to $q \overset{a}{\longrightarrow} \widehat{\mu}$ as a *transition* in NFSs. Intuitively, such a transition says that in state q and with the occurrence of event a, the set of all the allowable possibility degrees of entering q' is $\widehat{\mu}(q')$. An NFS differs from an NFDES in that possibility degrees of entering next states are sets rather than single values. It follows from definition that each NFDES can be viewed as an NFS in the obvious way.

In practice, one may not need the freedom of ascribing arbitrary sets of allowable possibility degrees to transitions. A special case is to restrict the sets to intervals in the form of $[\alpha, \beta]$, $(\alpha, \beta]$, $[\alpha, \beta)$, or (α, β) for $0 \le \alpha \le \beta \le 1$. Intuitively, interval specifications give bounds on the possibility degree of entering next states after performing a certain transition from a given state. The following is an example of NFSs using interval specifications.

Example 2 On the right side of Fig. 3, we show graphically an NFS $S = (\{P, F, E\}, \{a\}, \widehat{\delta}, P)$, where $\widehat{\delta}$ is defined as follows: $\widehat{\delta}(P, a) = \{\widehat{\mu}_1\}$ with $\widehat{\mu}_1(P) = [0, 0.5]$, $\widehat{\mu}_1(F) = [0.1, 1]$, $\widehat{\mu}_1(E) = [0, 1]$; $\widehat{\delta}(F, a) = \{\widehat{\mu}_2\}$ with $\widehat{\mu}_2(P) = [0, 0.3]$, $\widehat{\mu}_2(F) = [0.2, 1]$, $\widehat{\mu}_2(E) = [0, 1]$; $\widehat{\delta}(E, a) = \emptyset$. The intended meaning of the states is the same as that in Introduction and the unique event a represents an antibiotic drug. The NFS S is intended to specify the permitted reaction to the drug; for instance, the possibility degree that a patient's condition being in poor state reverts to the state after using the drug is required by not greater than 0.5.

Fig. 3 An implementation G and a specification S

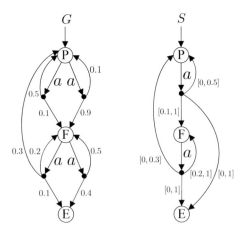

We are ready to introduce a satisfaction relation between NFDESs and NFSs which is based upon the idea that the possibility degrees arising in transitions of an NFDES must be in the set of possibility degrees arising in the corresponding transitions of an NFS. To state the definition, we need the following notion, which is used but not named in [33]. For any relation $R \subseteq X \times Y$, a pair (U, V) with $U \subseteq X$ and $V \subseteq Y$ is called R-*correlational* if $\{(x, y) \in R \mid x \in U\} = \{(x, y) \in R \mid y \in V\}$. It is easy to check that if $X = Y$ and R is an equivalence relation, then (U, V) is R-correlational if and only if $U = V = \cup_i C_i$ for some equivalence classes $C_i \in X/R$. Given $\mu \in \mathscr{F}(X)$ and $\eta \in \mathscr{F}(Y)$, we say that μ and η are *equivalent* with respect to R, denoted $\mu \equiv_R \eta$, if $\mu(U) = \eta(V)$ for every R-correlational pair (U, V). In addition, for any $\widehat{\mu} \in \widehat{\mathscr{F}}(Q)$ and $\eta \in \mathscr{F}(Q)$, we say that η is an *instance* of $\widehat{\mu}$, denoted $\eta \models \widehat{\mu}$, if $\eta(q) \in \widehat{\mu}(q)$ for every $q \in Q$.

Definition 7 Let $G = (Q, \Sigma, \delta, q_0)$ be an NFDES and $S = (Q', \Sigma, \widehat{\delta}, q_0')$ be an NFS. A relation $R \subseteq Q \times Q'$ is called a satisfaction relation between G and S if for all $(p, q) \in R$ and $a \in \Sigma$, whenever $p \xrightarrow{a} \mu$, there exist $q \xrightarrow{a} \widehat{\eta}$ and $\eta \models \widehat{\eta}$ such that $\mu \equiv_R \eta$. We say that p implements q, denoted $p \sqsubseteq q$, if $(p, q) \in R$ for some satisfaction relation R, and say that G implements S, denoted $G \sqsubseteq S$, if $q_0 \sqsubseteq q_0'$.

Intuitively, if $G \sqsubseteq S$, then it is possible for S to simulate, or "track", the behavior of G. In other words, the behavior of an implementation G is allowed by the specification S. Note that it does not require G to simulate the behavior of S.

Example 3 To illustrate further the definition of satisfaction relation, let us consider the NFDES G and the NFS S in Fig. 3. We may simply take $R = \{(P, P), (F, F), (E, E)\}$, which relates states with the same label to each other. Keep the notation in Example 2. For any $P \xrightarrow{a} \mu$ in G, we have that $P \xrightarrow{a} \widehat{\mu}_1$ in S. Taking $\eta = \mu$, we see that $\eta \models \widehat{\mu}_1$, and it is obvious that $\mu(U) = \eta(V)$ for every R-correlational pair (U, V). Similarly, one can consider the transitions leaving from the state F. Consequently, the state P in G implements the corresponding state in S, and thus, $G \sqsubseteq S$.

The following observation indicates that the class of satisfaction relations is closed under arbitrary unions.

Proposition 4 Let $G = (Q, \Sigma, \delta, q_0)$ be an NFDES and $S = (Q', \Sigma, \widehat{\delta}, q_0')$ be an NFS. If for any $i \in I$, R_i is a satisfaction relation between G and S, then so is $\cup_{i \in I} R_i$.

Proof It is routine by using [2, Lemma 1], so we do not go into the details.

We thus see that the union of all satisfaction relations gives rise to the largest satisfaction relation.

The next theorem implies that bisimilar NFDESs can implement the same NFS, that is, if $G_1 \sim G_2$ and $G_2 \sqsubseteq S$, then $G_1 \sqsubseteq S$.

Theorem 2 Let $G = (Q, \Sigma, \delta, q_0)$ be an NFDES and let $S = (Q', \Sigma, \widehat{\delta}, q_0')$ be an NFS. For any $p, q \in Q$ and $r \in Q'$, if $p \sim q$ and $q \sqsubseteq r$, then $p \sqsubseteq r$.

Proof As $p \sim q$, there exists a bisimulation R_1 on Q that contains (p, q). Therefore, for any $p \xrightarrow{a} \mu$, there is $q \xrightarrow{a} \eta$ such that $\mu(C) = \eta(C)$ for any $C \in Q/R_1$, which implies that $\mu \equiv_{R_1} \eta$. It follows from $q \sqsubseteq r$ that there is $R_2 \subseteq Q \times Q'$ such that $(q, r) \in R_2$, and moreover, for any $q \xrightarrow{a} \eta$, there are $r \xrightarrow{a} \widehat{\theta}$ and $\theta \models \widehat{\theta}$ with $\eta \equiv_{R_2} \theta$. Consider $R = R_1 \circ R_2 = \{(p', r') \mid \exists q' \in Q, (p', q') \in R_1, (q', r') \in R_2\} \subseteq Q \times Q'$. To show that R is a satisfaction relation between G and S, we need to verify that for any $(p', r') \in R$ and $a \in \Sigma$, if $p' \xrightarrow{a} \mu$, then there exist $r' \xrightarrow{a} \widehat{\theta}$ and $\theta \models \widehat{\theta}$ with $\mu \equiv_R \theta$. This is a fairly routine exercise, so we omit the details here.

6   Conclusion and Future Work

In the paper, we have presented a model for FDESs with nondeterminism, thus extending existing models. Bisimulation has been introduced for NFDESs as a finer behavioral measure. We have examined some properties of bisimulation and in particular, proven that bisimilarity is preserved by the product and parallel composition operators. Moreover, we have proposed the concept of NFSs and introduced a satisfaction relation between NFDESs and NFSs.

There are some problems which arise from the present formalization and are worth further studying. Firstly, it is desirable to use bisimulation equivalence to reduce an NFDES without losing essential information about its behavior. Secondly, the refinement of NFSs, which corresponds to restricting or specifying more tightly the possible behavior, remains an interesting problem. Or, one may "approximate" a given NFDES by an NFS in the sense that there is a satisfaction relation between the given system and the approximating system. Then checking certain properties for the approximating system will also guarantee these properties for the given system. Finally, the supervisory control theory based on bisimulation for NFDESs, which may be an extension of the related work [37, 38] for classical nondeterministic DESs, is yet to be established.

Acknowledgements This work was supported in part by the National Natural Science Foundation of China under Grants 61672023, 61772035, and 61751210, Guangxi Natural Science Foundation of China under Grant 2018GXNSFAA281326, and Guangxi Key Laboratory of Trusted Software under Grant kx201911.

References

1. Benmessahel, B., Touahria, M., Nouioua, F., Gaber, J., Lorenz, P.: Decentralized prognosis of fuzzy discrete-event systems. Iran. J. Fuzzy Syst. **16**(3), 127–143 (2019)
2. Cao, Y., Chen, G., Kerre, E.: Bisimulations for fuzzy-transition systems. IEEE Trans. Fuzzy Syst. **19**(3), 540–552 (2011)
3. Cao, Y., Ezawa, Y.: Nondeterministic fuzzy automata. Inform. Sci. **191**, 86–97 (2012)

4. Cao, Y., Sun, S.X., Wang, H., Chen, G.: A behavioral distance for fuzzy transition systems. IEEE Trans. Fuzzy Syst. **21**(4), 735–747 (2013)
5. Cao, Y., Ying, M.: Supervisory control of fuzzy discrete event systems. IEEE Trans. Syst. Man Cybern. B Cybern. **35**(2), 366–371 (2005)
6. Cao, Y., Ying, M.: Observability and decentralized control of fuzzy discrete-event systems. IEEE Trans. Fuzzy Syst. **14**(2), 202–216 (2006)
7. Cao, Y., Ying, M., Chen, G.: State-based control of fuzzy discrete-event systems. IEEE Trans. Syst. Man Cybern. B Cybern. **37**(2), 410–424 (2007)
8. Chen, T., Han, T., Cao, Y.: Polynomial-time algorithms for computing distances of fuzzy transition systems. Theor. Comput. Sci. **727**, 24–36 (2018)
9. Ćirić, M., Stamenković, A., Ignjatović, J., Petković, T.: Fuzzy relation equations and reduction of fuzzy automata. J. Comput. Syst. Sci. **76**(7), 609–633 (2010)
10. Deng, W., Qiu, D.: State-based decentralized diagnosis of bi-fuzzy discrete event systems. IEEE Trans. Fuzzy Syst. **25**(4), 3854–867 (2017)
11. Du, X., Ying, H., Lin, F.: Theory of extended fuzzy discrete-event systems for handling ranges of knowledge uncertainties and subjectivity. IEEE Trans. Fuzzy Syst. **17**(2), 316–328 (2009)
12. Garg, V.K., Kumar, R., Marcus, S.I.: A probabilistic language formalism for stochastic discrete-event systems. IEEE Trans. Automat. Contr. **44**(2), 280–293 (1999)
13. Heymann, M., Lin, F.: Discrete-event control of nondeterministic systems. IEEE Trans. Automat. Contr. **43**(1), 3–17 (1998)
14. Hoare, C.A.R.: Communicating Sequential Processes. Prentice-Hall, Englewood Cliffs, NJ (1985)
15. Huq, R., Mann, G.K.I., Gosine, R.G.: Behavior-modulation technique in mobile robotics using fuzzy discrete event system. IEEE Trans. Robot. **22**(5), 903–916 (2006)
16. Jiang, S., Kumar, R.: Supervisory control of nondeterministic discrete-event systems with driven events via masked prioritized synchronization. IEEE Trans. Automat. Contr. **47**(9), 1438–1449 (2002)
17. Kumar, R., Garg, V.K.: Control of stochastic discrete event systems modeled by probabilistic languages. IEEE Trans. Automat. Contr. **46**(4), 593–606 (2001)
18. Larsen, K.G., Skou, A.: Bisimulation through probabilistic testing. Inform. Comput. **94**, 1–28 (1991)
19. Lawford, M., Wonham, W.M.: Supervisory control of probabilistic discrete event systems. In: Proceedings of 36th Midwest Symposium on Circuits and Systems, vol. 1, pp. 327–331. IEEE, Detroit, MI (1993)
20. Li, Y.H., Lin, F., Lin, Z.H.: Supervisory control of probabilistic discrete-event systems with recovery. IEEE Trans. Automat. Contr. **44**(10), 1971–1975 (1999)
21. Lin, F., Ying, H.: Modeling and control of fuzzy discrete event systems. IEEE Trans. Syst. Man Cybern. B Cybern. **32**(4), 408–415 (2002)
22. Lin, F., Ying, H.: State-feedback control of fuzzy discrete-event systems. IEEE Trans. Syst. Man Cybern. B Cybern. **40**(3), 951–956 (2010)
23. Lin, F., Ying, H., MacArthur, R.D., Cohn, J.A., Barth-Jones, D., Crane, L.R.: Decision making in fuzzy discrete event systems. Inform. Sci. **177**(18), 3749–3763 (2007)
24. Liu, J.P., Li, Y.M.: The relationship of controllability between classical and fuzzy discrete-event systems. Inform. Sci. **178**(21), 4142–4151 (2008)
25. Liu, R., Wang, Y.X., Zhang, L.: An FDES-based shared control method for asynchronous brain-actuated robot. IEEE Trans. Cybern. **46**(6), 1452–1462 (2016)
26. Milner, R.: Communication and Concurrency. Prentice-Hall, Englewood Cliffs, New Jersey (1989)
27. Pantelic, V., Postma, S.M., Lawford, M.: Probabilistic supervisory control of probabilistic discrete event systems. IEEE Trans. Automat. Contr. **54**(8), 2013–2018 (2009)
28. Park, D.: Concurrency and automata on infinite sequences. In: Proceedings of 5th GI-Conference Theoretical Computer Science. Lecture Notes in Computer Science, vol. 104, pp. 167–183. Springer (1981)

29. Petković, T.: Congruences and homomorphisms of fuzzy automata. Fuzzy Sets Syst. **157**, 444–458 (2006)
30. Qiu, D.W.: Supervisory control of fuzzy discrete event systems: a formal approach. IEEE Trans. Syst. Man Cybern. B Cybern. **35**(1), 72–88 (2005)
31. Qiu, D.W., Liu, F.C.: Fuzzy discrete-event systems under fuzzy observability and a test algorithm. IEEE Trans. Fuzzy Syst. **17**(3), 578–589 (2009)
32. Sun, D.D., Li, Y.M., Yang, W.W.: Bisimulation relations for fuzzy finite automata. Fuzzy Syst. Math. **23**, 92–100 (2009). (in Chinese)
33. de Vink, E.P., Rutten, J.J.M.M.: Bisimulation for probabilistic transition systems: a coalgebraic approach. Theor. Comput. Sci. **221**, 271–293 (1999)
34. Wonham, W.M.: Supervisory control of discrete-event systems. University of Toronto, Toronto, ON, Canada, Technical Report (2005). http://www.control.utoronto.ca/DES/
35. Ying, H., Lin, F., MacArthur, R.D., Cohn, J.A., Barth-Jones, D.C., Ye, H., Crane, L.R.: A fuzzy discrete event system approach to determining optimal HIV/AIDS treatment regimens. IEEE Trans. Inform. Tech. Biomed. **10**(4), 663–676 (2006)
36. Ying, H., Lin, F., MacArthur, R.D., Cohn, J.A., Barth-Jones, D.C., Ye, H., and L. R. Crane, "A self-learning fuzzy discrete event system for HIV/AIDS treatment regimen selection. IEEE Trans. Syst. Man Cybern. B Cybern. **37**(4), 966–979 (2007)
37. Zhou, C., Kumar, R., Jiang, S.: Control of nondeterministic discrete-event systems for bisimulation equivalence. IEEE Trans. Automat. Contr. **51**(5), 754–765 (2006)
38. Zhou, C., Kumar, R.: Control of nondeterministic discrete event systems for simulation equivalence. IEEE Trans. Automat. Sci. Eng. **4**(3), 340–349 (2007)

Italian Folk Multiplication Algorithm Is Indeed Better: It Is More Parallelizable

Martine Ceberio, Olga Kosheleva and Vladik Kreinovich

Abstract Traditionally, many ethnic groups had their own versions of arithmetic algorithms. Nowadays, most of these algorithms are studied mostly as pedagogical curiosities, as an interesting way to make arithmetic more exciting to the kids: by applying to their patriotic feelings—if they are studying the algorithms traditionally used by their ethnic group—or simply to their sense of curiosity. Somewhat surprisingly, we show that one of these algorithms—a traditional Italian multiplication algorithm—is actually in some reasonable sense better than the algorithm that we all normally use—namely, it is easier to parallelize.

1 Formulation of the Problem

How we learn to multiply numbers. How students learn multiplication is school?

- First, they memorize the multiplication table—which enables them to multiply 1-digit numbers.
- Then, they learn how to multiply a multi-digit number by a digit.
- Finally, they learn how to multiply two multi-digit numbers.

Let us recall how this is taught in school.
To multiply a multi-digit number by a digit, e.g., multiply 23 by 4,

M. Ceberio · O. Kosheleva · V. Kreinovich (✉)
University of Texas at El Paso, El Paso, TX 79968, USA
e-mail: vladik@utep.edu

M. Ceberio
e-mail: mceberio@utep.edu

O. Kosheleva
e-mail: olgak@utep.edu

© Springer Nature Switzerland AG 2020
M. Ceberio and V. Kreinovich (eds.), *Decision Making under Constraints*,
Studies in Systems, Decision and Control 276,
https://doi.org/10.1007/978-3-030-40814-5_7

```
   23
X  4
---
    ?
```

we start with the lowest digit—in this case, with 3, and multiply it by 4. From the multiplication table, we know that the result is 12, so we place 2 in the corresponding digit of a product, and remember 1 as a *carry*, to be added to the next digit:

```
   23
X  4
---
  ?2
```

Then, we multiply the next digit (in this case, 2) by 4, getting 8, and add the carry (in this case, 1) to this product, getting 9:

```
   23
X  4
---
  92
```

Similarly, if we multiply 23 by 6, we:

- get $3 \cdot 6 = 18$, so the carry is 1, and
- then compute $2 \cdot 6 + 1 = 13$,

so the result is:

```
   23
X  6
---
 138
```

Once the students master the art of multiplying a multi-digit number by a digit, they learn how to multiply two multi-digit numbers:

- first, we multiply the first number by each digit of the second number, and
- then, we add up all the resulting products.

For example, to multiply 23 by 64, we first perform the above two multiplications, and then add the results:

```
     23
X   64
----
    92
+ 138
------
  1472
```

Ethnic multiplication algorithms. In the past, different ethic groups used different algorithms for multiplication. Probably the most well known is the Russian multiplication algorithm (see, e.g., [1, 3, 5]), in which to compute the product $a \cdot b$, we, in effect:

- translate the second number b into the binary code, i.e., represent it as

$$b = 2^{i_1} + 2^{i_2} + \cdots + 2^{i_k}$$

for some $i_1 > i_2 > \cdots > i_k$, then
- consequently double the first number a, to get the values

$$a, \ 2^1 \cdot a, \ 2^2 \cdot a, \ \ldots, \ 2^{i_1} \cdot a,$$

- and after this, add the products corresponding to the powers of 2 that form b:

$$a \cdot b = 2^{i_1} \cdot a + 2^{i_2} \cdot a + \cdots + 2^{i_k} \cdot a.$$

If we only have two numbers a and b to multiply, the Russian multiplication algorithm seems to require a lot of unnecessary steps, but it starts making sense if we have to multiply the same number a by different values b. For example—and this is where this algorithm originated—a merchant is selling some material by yards, and he (in the old days, it was usually he) needs to find the price of different amounts of material. In this case, a—the price per yard—remains the same, while the length b changes. The advantage is that in this case, we perform all the doublings only once—as result, we only need:

- to translate into binary code—and for this, it is sufficient to divide by 2, and then
- to add the corresponding products $2^i \cdot a$.

Different ethnic groups had different algorithms. For example, in the traditional Italian folks multiplication algorithm (see, e.g., [4] and references therein), we:

- multiply each digit of the first number by each digit of the second number, and then
- add the results.

For example, in this algorithm, the multiplication of 23 by 64 takes the following form:

```
   23
X  64
 ----
   12   = 3 x 4
   18   = 3 x 6
    8   = 2 x 4
```

```
  +  12
  -------
    1472
```

How are ethnic algorithms viewed now. At present, the ethnic algorithms are studied mostly by historians of science and by pedagogues. To pedagogues, such algorithms are an interesting way to make arithmetic more exciting to the kids.

In general, studying different algorithms raises the students' curiosity level. Also, studying algorithms of one's own ethnic group is enhanced by the students' patriotic feelings—although, strangely enough, OK and VK, when studying arithmetic in Russia, never heard of the Russian multiplication algorithm.

What we show in this paper. Our goal is to show that ethnic algorithms actually make sense—many of them are, in some sense, better than the algorithm that we learn at school. We have already mentioned this for the Russian multiplication algorithm.

In this short paper, we show that the Italian folk multiplication algorithm also has its advantages over the traditional modern-school multiplication—namely, the Italian algorithm is easier to parallelize.

2 Italian Algorithm Is Better: An Explanation

Why parallelization. Nowadays, most multiplication is performed by computers, and computers have no problem multiplying large numbers. However, in the past, multiplication was not easy. In the Middle Ages, when even literacy was rather an exception, those who could multiply never needed to do a back-breaking menial work: they could easily find employment as assistants to merchants.

If one needs to multiply two large numbers, and the result is important—a natural idea is to ask for help, to divide the job, so that two specialists in this complex art of multiplication could perform some operations at the same time ("in parallel") and thus, speed up the process.

In a nutshell, this is the same reason why modern computer-based computations use parallelization: if a computation takes too long on a single processor, a reasonable idea is to have several processors working in parallel.

Which algorithm is easier to parallelize. From this viewpoint, it is desirable to check which of the two algorithms—the usual one or the Italian folk one—is easier to parallelize.

How do we gauge easiness? Each of the two algorithms consists of two stages:

- the first, multiplication stage, and
- the second stage, in which add the multiplication results.

Clearly, addition is much easier than multiplication. From this viewpoint, when we talk about parallelization, we should emphasize the need to parallelize the first (multiplication) part of each algorithm.

What can be parallelized in the traditional multiplication algorithm. In the traditional multiplication algorithm, to compute $a \cdot b$, we multiply a by each of the digits of b, and then add the resulting products. Multiplication of a by each of the b's digits does not depend on the multiplication on any other digit, so all these multiplications can be performed in parallel.

In the above example:

- one person can multiply 23 by 4, getting 92, while
- at the same time, another person could multiply 23 by 6, resulting in 138, after which they can easily add the results.

However, no further parallelization is possible (unless we modify the algorithm). Namely, the way a multi-digit number is multiplied is sequential:

- we do not get the second-from-last digit of the product until we have computed the last digit,
- we do not get the third-from-last digit of the product until we have computed the second-form-last digit, etc.

What can be parallelized in the Italian folk multiplication algorithm. In contrast, in the traditional Italian algorithm, when we first multiply each digit of the first number by each digit of the second number, all these multiplications can be done in parallel.

For example, when we multiply 23 by 64:

- the first person multiplies 3 by 4,
- the second person multiplies 3 by 6,
- the third person multiplies 2 by 4, and
- the fourth person multiplies 2 by 6.

All these four multiplications can be performed at the same time—i.e., in parallel—after which all that remains is an easy task of *adding* all four multiplication results.

Conclusion. We see that the Italian algorithm is indeed better than the traditional one—in the sense that it is easier to parallelize than the traditional multiplication algorithm.

Caution. The above arguments make sense to us, but the readers should be warned that, while these arguments seem reasonable, they do not work if we consider a traditional computer science approach to algorithm complexity—which is based on considering the length of the inputs tending to infinity; see, e.g., [2].

Indeed, as the number of digits B in the second number b increases, it becomes much larger than 10. In this case, in the traditional multiplication algorithm, we no longer need to perform B multiplication of a by a 1-digit number: it is sufficient to find the product of a by each of the 10 digits, and then simply place the corresponding product into the resulting sum. (To be more precise, we need 8 multiplications, since multiplying by 0 and 1 is trivial.)

This observation makes the traditional algorithm somewhat easier—but still, multiplying a very long number by a digit is, in the traditional algorithm, not naturally parallelizable.

Acknowledgements This work was supported in part by the US National Science Foundation grant HRD-1242122.

References

1. Bunt, L.N.H., Jones, P.S., Bedient, J.D.: The Historical Roots of Elementary Mathematics. Dover, New York (1988)
2. Cormen, T.H., Leiserson, C.E., Rivest, R.L., Stein, C.: Introduction to Algorithms. MIT Press, Cambridge, Massachusetts (2009)
3. Knuth, D.E.: The Art of Computer Programming, vol. 2. Seminumerical Algorithms, Addison Wesley, Reading, Massachusetts (1969)
4. Llorente, A.: 3 sencillos métodos para aprender a multiplicar sin calculadora, BBC Mundo, posted Nov 22 (2017). http://www.bbc.com/mundo/noticias-42020116
5. Vargas, J.I., Kosheleva, O.: Russian peasant multiplication algorithm, RSA cryptosystem, and a new explanation of half-orders of magnitude. J. Uncertain Syst. **1**(3), 178–184 (2007)

Reverse Mathematics Is Computable for Interval Computations

Martine Ceberio, Olga Kosheleva and Vladik Kreinovich

Abstract For systems of equations and/or inequalities under interval uncertainty, interval computations usually provide us with a box whose all points satisfy this system. Reverse mathematics means finding necessary and sufficient conditions, i.e., in this case, describing the set of *all* the points that satisfy the given system. In this paper, we show that while we cannot always exactly describe this set, it is possible to have a general algorithm that, given $\varepsilon > 0$, provides an ε-approximation to the desired solution set.

1 Formulation of the Problem

What is reverse mathematics. In mathematics, whenever a new theorem is proven, often, it later turns out that this same conclusion can be proven under weaker conditions.

For example, first, it was proven that if for a continuous function $f(x)$ from real numbers to real numbers, we have $f(a + b) = f(a) + f(b)$ for all a and b, then this function $f(x)$ is linear, i.e., $f(a) = k \cdot a$ for some k, Later on, it turned out that the same is true not only for continuous functions, but also for all measurable functions.

Because of this phenomenon, every time a new result is proven, researchers start analyzing whether this result can be proven under weaker conditions. In the past, usually, weaker and weaker conditions were found. Lately, however, in some problems, it has become possible to find the weakest possible conditions under which the given conclusion is true.

M. Ceberio · O. Kosheleva · V. Kreinovich (✉)
University of Texas at El Paso, El Paso, TX 79968, USA
e-mail: vladik@utep.edu

M. Ceberio
e-mail: mceberio@utep.edu

O. Kosheleva
e-mail: olgak@utep.edu

© Springer Nature Switzerland AG 2020
M. Ceberio and V. Kreinovich (eds.), *Decision Making under Constraints*,
Studies in Systems, Decision and Control 276,
https://doi.org/10.1007/978-3-030-40814-5_8

65

From the logical viewpoint, the fact that the condition A in the implication $A \Rightarrow B$ cannot be weakened means that A is equivalent to B, i.e., that also $B \Rightarrow A$—in other words, that we can *reverse* the implication. Because of this, the search for such weakest possible condition is known as the *reverse mathematics*; see, e.g., [6].

What does reverse mathematics mean for interval computations? The main problem of interval computations (see, e.g., [3–5]) is, given an algorithm $f(x_1, \ldots, x_n)$ and ranges $[\underline{x}_i, \overline{x}_i]$, to find the range $[\underline{y}, \overline{y}]$ of possible values of $y = f(x_1, \ldots, x_n)$ when $x_i \in [\underline{x}_i, \overline{x}_i]$ for all i.

In practice, this can be used, e.g., to check that under all possible values of the parameters x_i from the corresponding intervals, the system is stable, when stability is described by an inequality $f(x_1, \ldots, x_n) \leq y_0$ for some value y_0. Once we know the range, this checking is equivalent to simply checking whether $\overline{y} \leq y_0$.

Similarly, if $f(x_1, \ldots, x_n)$ is the amount of potentially polluting chemical released by a plant under conditions x_i, checking whether the level of this chemical never exceeds the desired threshold y_0 is also equivalent to simply checking whether $\overline{y} \leq y_0$.

In addition to knowing that $x_i \in [\underline{x}_i, \overline{x}_i]$, we often have additional constraints on the values x_i, which make the problem more complex.

We may also need to check more complex conditions. For example, in solving system of equations under interval uncertainty, we are often interested in finding all the values $x = (x_1, \ldots, x_n)$ for which, for all possible values $a_1, \ldots, a_m$ from the corresponding intervals, there exist appropriate controls $c_1, \ldots, c_p$ from the given intervals for which a desired inequality $f(x, a, c) \leq y_0$ holds. Since all physical quantities are bounded, we can safely assume that all variables in the quantifiers are bounded.

In general, we have a property $P(x_1, \ldots, x_n)$ which can be either a simple inequality like $f(x_1, \ldots, x_n) \leq y_0$, or it can be a complex formula obtained from simply inequalities by using logical connectives "and" (&) "or" ($\vee$), and "not" ($\neg$), and quantifiers $\forall x$ and $\exists x$ over real numbers. By using interval methods, we find a box $B = [\underline{x}_1, \overline{x}_1] \times \ldots \times [\underline{x}_n, \overline{x}_n]$ for which the desired property $P(x)$ holds for all points $x \in B$. In this context, reverse mathematics means trying to find not just this box, but also the whole set of all the tuples x for which the property $P(x)$ holds.

What we do in this paper. In this paper, we show that such a set can be indeed computed—maybe not exactly, but at least with any possible accuracy.

2 Definitions and the Main Result

Computable numbers, sets, etc.: reminder. According to computable mathematics (see, e.g., [1, 7]), a real number x is *computable* if there exists an algorithm that, given a natural number n, generates a rational number r_n for which $|r_n - x| \leq 2^{-n}$. A tuple of computable numbers is called a *computable tuple*.

A bounded set S is called *computable* if there exists an algorithm that, given a natural number n, generates a finite list S_n of computable tuples for which $d_H(S, S_n) \le 2^{-n}$, where $d_H(A, B)$ is the Hausdorff distance: the smallest $\varepsilon > 0$ for which:

- every element $a \in A$ is ε-close to some element $b \in B$, and
- every element $b \in B$ is ε-close to some element $a \in A$.

A *computable function* is a function $f(x_1, \ldots, x_n)$ for which two algorithms exist:

- the main algorithm that, given rational values $r_1, \ldots, r_n$, returns a computable number $f(r_1, \ldots, r_n)$, and
- an auxiliary algorithm that, given a rational number $\varepsilon > 0$, returns a rational number $\delta > 0$ for which $d(x, x') \le \delta$ implies $d(f(x), f(x')) \le \varepsilon$.

Most arithmetic and elementary functions are everywhere computable. (The only exceptions are discontinuous functions like sign or tangent.)

It is known (and it can be easily proven):

- that min and max are computable,
- that composition of two computable functions is computable, and
- that the maximum and minimum of a computable function over a computable set are also computable.

It is also known that for every computable function f on a computable set S, and for every two values $y^- < y^+$ for which $\min_{x \in S} f(x) < y^-$, there exists a value $y_0 \in [y^-, y^+]$ for which the set $\{x : f(x) \le y_0\}$ is computable [1].

There are also known negative results: e.g.,

- that it is not possible, given two computable numbers x and x', to check whether $x \le x'$, and,
- as a consequence, that it is, in general, not possible, given a computable function f and a number y, to produce a computable set $\{x : f(x) \le y_0\}$—otherwise, for a constant $f(x) = c$, we would get an algorithm for checking whether $c \le y_0$.

Definition 1 Let $v_1, \ldots$ be real-valued variables. For each of these variables, we have bounds $\underline{V}_i \le v_i \le \overline{V}_i$.

- By a *term*, we mean an expression of the type $f(v_{i_1}, \ldots, v_{i_m})$, where f is a computable function and v_i are given variables.
- By an *elementary formula*, we means an expression of one of the types $t_1 < t_2, t_1 \le t_2$, or $t_1 = t_2$, where t_1 and t_2 are terms.
- By a *property* $P(x_1, \ldots, x_n)$, we mean any formula with free variables $x_1, \ldots, x_n$ which is obtained from elementary formulas by using logical connectives $\&$, $\vee$, $\neg$, and quantifiers $\forall v_{i_{\in[\underline{V}_i, \overline{V}_i]}}$ and $\exists v_{i_{\in[\underline{V}_i, \overline{V}_i]}}$.

Comment. To simplify the further description, let us represent each equality $t_1 = t_2$ as two inequalities $t_1 \le t_2$ and $t_2 \le t_1$.

Definition 2 Let $\varepsilon > 0$ be a real number.

- We say that elementary formulas $t_1 \leq t_2$ (or $t_1 < t_2$) and $t_1 \leq t_2 + \varepsilon'$ (or $t_1 < t_2 + \varepsilon'$) are ε-*close* if $|\varepsilon'| \leq \varepsilon$.
- We say that the formulas $P(x_1, \ldots, x_n)$ and $P(x_1', \ldots, x_n')$ are ε-*close* if P' is the result of replacing, in the formula P, each elementary formula with an ε-close one.

Comment. In practice, all the values are measured with some accuracy. Thus, if ε is sufficiently small, the two ε-close elementary formulas are practically indistinguishable—and thus, in general, ε-close properties are indistinguishable as well.

Proposition 1 *Let $P(x_1, \ldots, x_n)$ be a property which is satisfied for all the tuples x from a given box. Then, based on the property and the box, we can compute the set $\{x : P'(x)\}$ for some property P' which is ε-close to P.*

Comment. This result can be further strengthened.

Proposition 2 *Let $P(x_1, \ldots, x_n)$ be a property which is satisfied for all the tuples x from a given box. Then, based on the property and the box, we can compute the set $S = \{x : P'(x)\}$ for some property P' which is ε-close to P and for which $S'' \stackrel{\text{def}}{=} \{x : P''(x)\} \subseteq S$ for all properties P'' which are $(\varepsilon/2)$-close to P.*

3 Proof

Proof of Proposition 1. First, let us transform the original property $P(x)$ into a prenex normal form, i.e., into the form (see, e.g., [2]) in which we first have quantities, and then the quantifier-free part. Indeed, if we have a logical connective outside quantifires, we can move the quantifier out by using the equivalent transformations $\neg \forall x\, P \to \exists x\, \neg P$, $\forall x\, P \vee Q \to \forall x (P \vee Q)$, $\forall x\, P \,\&\, Q \to \forall x (P \,\&\, Q)$, $\neg \exists x\, P \to \forall x\, \neg P$, $\exists x\, P \vee Q \to \exists x (P \vee Q)$, and $\exists x\, P \,\&\, Q \to \exists x (P \,\&\, Q)$.

Then, we can use de Morgan rules

$$\neg(A \,\&\, B) \to (\neg A) \vee (\neg B) \text{ and } \neg(A \vee B) \to (\neg A) \,\&\, (\neg B)$$

to move all negations inside. When applied to an elementary formulas $t_1 \leq t_2$ or $t_1 < t_2$, negation simply means a change in the inequality sign: $\neg(t_1 \leq t_2) \to t_2 < t_1$ and $\neg(t_1 < t_2) \to t_2 \leq t_1$.

In the resulting formula, let us replace all $<$ with $\leq$; this will not change ε-closeness. Let us now describe the resulting property P_0 in the equivalent form $F(x_1, \ldots, x_n) \leq 0$, for some computable function F.

- Each elementary formula $t_1 \leq t_2$ can be equivalently reformulated as

$$t_1 - t_2 \leq 0.$$

- Each formula $(F_1 \leq 0) \vee (F_2 \leq 0)$ can be equivalently reformulated as

$$\min(F_1, F_2) \leq 0.$$

- Each formula $(F_1 \leq 0) \,\&\, (F_2 \leq 0)$ can be equivalently reformulated as

$$\max(F_1, F_2) \leq 0.$$

- Each formula $\exists v_{i \in [\underline{V}_i, \overline{V}_i]}\, F(v_i, \ldots) \leq 0$ can be equivalently reformulated as

$$\min_{v_i \in [\underline{V}_i, \overline{V}_i]}\, F(v_i, \ldots) \leq 0.$$

- Each formula $\forall v_{i \in [\underline{V}_i, \overline{V}_i]}\, F(v_i, \ldots) \leq 0$ can be equivalently reformulated as

$$\max_{v_i \in [\underline{V}_i, \overline{V}_i]}\, F(v_i, \ldots) \leq 0.$$

For the resulting function $F(x_1, \ldots, x_n)$, for $y^- = 0$ and $y^+ = \varepsilon$, there exists a number $\varepsilon_0 \in (0, \varepsilon)$ for which the set $S_0 \stackrel{\text{def}}{=} \{x : F(x) \leq \varepsilon_0\}$ is computable.

The corresponding inequality $F(x) \leq \varepsilon_0$ is equivalent to $F'(x) \leq 0$, where $F'(x) \stackrel{\text{def}}{=} F(x) - \varepsilon_0$. This inequality can be obtained if we replace, in the formula P_0, each elementary formula $t_1 \leq t_2$ with a formula $t_1 \leq t_2 + \varepsilon_0$. Since $\varepsilon_0 < \varepsilon$, this transformation keeps all elementary formulas ε-close to the original ones. So, the resulting formula P'_0 is ε-close to the formula P_0 and we have $S_0 = \{x : P'_0(x)\}$.

When we went from P to P_0, all we did was changed the sign of some inequalities. This, in turn, can be obtained by appropriately changing the elementary formulas from the original property P to ε-close ones. Thus, indeed, the set S_0 can be represented as $S_0 = \{x : P'(x)\}$ for the resulting formula P' which is ε-close to P.

Proof of Proposition 2 is similar, except that instead of $y^- = 0$ we take $y^- = \varepsilon/2$. Then, for every property P'' which is $(\varepsilon/2)$-close to P, on each level of designing a function $F(x_1, \ldots, x_n)$, we will have $F \leq F'' + \varepsilon/2$ for the function F'' corresponding to the property P''. Thus, at the end, we conclude that $F \leq F'' + \varepsilon/2$ and, since now $\varepsilon/2 < \varepsilon_0$, we conclude that $F(x) \leq F''(x) + \varepsilon_0$ and $F'(x) = F(x) - \varepsilon_0 \leq F''(x)$. Thus, $F''(x) \leq 0$ implies that $F'(x) \leq 0$. So, indeed, $P'' \subseteq S'$. The proposition is proven.

Acknowledgements This work was supported in part by the US National Science Foundation grant HRD-1242122.

References

1. Bishop, E.: Foundations of Constructive Analysis. McGraw-Hill, New York (1967)
2. Hinman, P.: Fundamentals of Mathematical Logic. A. K. Peters, Natick, Massachusetts (2005)
3. Jaulin, L., Kiefer, M., Didrit, O., Walter, E.: Applied Interval Analysis, with Examples in Parameter and State Estimation, Robust Control, and Robotics. Springer, London (2001)
4. Mayer, G.: Interval Analysis and Automatic Result Verification. de Gruyter, Berlin (2017)
5. Moore, R.E., Kearfott, R.B., Cloud, M.J.: Introduction to Interval Analysis. SIAM, Philadelphia (2009)
6. Stillwell, J.: Reverse Mathematics: Proofs from the Inside Out. Princeton University Press, Princeton, New Jersey (2018)
7. Weihrauch, K.: Computable Analysis. Springer, Berlin (2000)

Generalized Ordinal Sum Constructions of t-norms on Bounded Lattices

Antonín Dvořák and Michal Holčapek

Abstract In this contribution, we propose an ordinal sum construction of t-norms on bounded lattices from a system of lattices (endowed by t-norms) whose index set forms a bounded lattice. The ordinal sum construction is determined by a lattice-based sum of bounded lattices and interior operators.

1 Introduction

The operations of t-norms and t-conorms on the unit interval [13, 19] are important operations used in fuzzy set theory and fuzzy logic and also in their applications. They serve as natural interpretations of operations of conjunction and disjunction, respectively. Because fuzzy logic soon started to use more general structures of truth values [10], and these structures fall under the concept of bounded lattices, it was quite natural to begin to study t-norms and t-conorms on bounded lattices [1, 5, 21].

Ordinal sums in the sense of Clifford [4] are very important constructions of t-norms and t-conorms on the unit interval. They also provide a basis of a well-known representation theorem for continuous t-norms (t-conorms) as ordinal sums of isomorphic images of Łukasiewicz and product t-norms (t-conorms) [13, 15]. Generalizations of ordinal sum constructions for t-norms and t-conorms on bounded lattices have been intensively studied [3, 9, 14, 17, 18].

Recently [6] we proposed a new and more general definition of an ordinal sum of t-norms and t-conorms on bounded lattices and showed that ordinal sums proposed

This research was partially supported from the ERDF/ESF project AI-Met4AI (No. CZ.02.1.01/0.0/0.0/17_049/0008414). The additional support was also provided by the Czech Science Foundation through the project of No.18-06915S.

A. Dvořák · M. Holčapek (✉)
Institute for Research and Applications of Fuzzy Modeling, University of Ostrava, CE IT4Innovations, 30. dubna 22, 701 03 Ostrava, Czech Republic
e-mail: Michal.Holcapek@osu.cz

A. Dvořák
e-mail: Antonin.Dvorak@osu.cz

in [3, 9] are special cases of our definition. We used the concept of lattice interior and closure operators [16], which allow us to pick a sublattice (or sublattices) of a given bounded lattice appropriate for our construction.

The ordinal sum in the sense of Birkhoff [2] is an important construction method of ordered structures (posets, lattices, etc.). In [8], a generalization of this construction was proposed with an index set formed by a bounded lattice. In a recent paper [7], ordinal sums of t-norms and t-conorms with these index sets were defined and investigated.

The aim of this paper is to extend further the approach proposed in [6], that is, ordinal sums of t-norms and t-conorms using lattice interior and closure operators, in the direction of lattice-based index sets of [7]. Moreover, we propose a different definition of general ordinal sum of bounded lattices than that in [7] (formulated purely in terms of lattices and their homomorphisms) that should clarify, at least partially, the assumptions on the family of bounded lattices used in [7]. Deep comparison of both approaches to the generalized ordinal sum of bounded lattices is a subject of our future research. Due to space limitations, we consider only ordinal sums of t-norms using interior operators in this paper. Ordinal sums of t-conorms using closure operators can be defined analogously.

The paper is structured as follows. In Sect. 2, we recall notions of a bounded lattice, lattice homomorphisms, congruences on bounded lattices and also t-norms on bounded lattices. Section 3 contains the construction of an ordinal sum of bounded lattices using a bounded lattice as an index set. We use a notion of a semi-congruence (Definition 3) in the definition of a P-family of bounded lattices suitable for the construction of the ordinal sum of bounded lattices (Theorem 2). The main results concerning a new ordinal sum construction of t-norms from a system of lattices endowed by t-norms and indexed by a bounded lattice using interior operators are presented in Sect. 4. Section 5 contains conclusions and directions of further research.

2 Preliminaries

Let $L = (L, \wedge, \vee)$ denote a lattice, i.e., L is an algebraic structure of type $(2, 2)$, where the operations of meet $\wedge$ and join $\vee$ satisfy the axioms of idempotency, commutativity, associativity, and absorbtion (for details, we refer to [12, 20]), and let $\leq$ denote the lattice order on L, which is determined by $x \leq y$ if $x \wedge y = x$ for any $x, y \in L$. A lattice L is said to be *bounded* if there exist two elements $0_L, 1_L \in L$ such that for all $x \in L$ it holds that $0_L \vee x = x$ and $1_L \wedge x = x$. We call 0_L and 1_L the bottom and top element, respectively, and write this bounded lattice as $(L, \wedge, \vee, 0_L, 1_L)$. A subset K of a lattice L is said to be a *sublattice of* L if K is itself a lattice with respect to the meet and join operations of L. If L is a bounded lattice and K is its sublattice then K is also bounded and it holds that $0_K = 0_L$ and $1_K = 1_L$.[1] A sublattice K of L is said to be *convex* if $x \leq z \leq y$ and $x, y \in K$ implies that $z \in K$, for all $x, y, z \in L$. A map $f : L_1 \to L_2$ is said to be a *homomorphism of lattices* (a *lattice*

---

[1]This is a consequence of the fact that 0_L and 1_L are nullary operations in the bounded lattice L.

homomorphism) if $f(x \wedge y) = f(x) \wedge f(y)$ and $f(x \vee y) = f(x) \vee f(y)$ for any $x, y \in L_1$. The closed subinterval $[a, b]$ of the lattice L is the sublattice

$$[a, b] = \{x \in L \mid a \le x \le b\}.$$

Similarly, the open subinterval (a, b) of L is defined as $(a, b) = \{x \in L \mid a < x < b\}$. Definitions of semi-open intervals $(a, b]$ and $[a, b)$ are obvious. In what follows, we assume that each index set P has a structure of a bounded lattice $P = (P, \wedge, \vee, 0_P, 1_P)$.[2] Let $\mathbf{2} = \{0, 1\}$ denote the two-element bounded lattice with the bottom element 0 and the top element 1. Further, we use P^* to denote the lexicographic product of lattices P and $\mathbf{2}$, i.e., $P^* = \{(p, i) : p \in P, i \in \mathbf{2}\}$ and the meet $\wedge^*$ and joint $\vee^*$ on P^* are defined as follows:

$$(p, i) \wedge^* (q, j) = \begin{cases} (p, i \wedge j), & p = q, \\ (p, i), & p < q, \\ (q, j), & q < p, \\ (p \wedge q, 1), & \text{otherwise,} \end{cases} \tag{1}$$

$$(p, i) \vee^* (q, j) = \begin{cases} (p, i \vee j), & p = q, \\ (q, j), & p < q, \\ (p, i), & q < p, \\ (p \vee q, 0), & \text{otherwise.} \end{cases} \tag{2}$$

for any $(p, i), (q, j) \in P^*$. Obviously, $0_{P^*} = (0_P, 0)$ and $1_{P^*} = (1_P, 1)$. An example of the lexicographic product of P and $\mathbf{2}$ can be seen in Fig. 1 (right). An equivalence $\equiv$ on L is said to be a *congruence* if, for any $x, x', y, y' \in L$ such that $x \equiv x'$ and $y \equiv y'$, we have

$$x \wedge y \equiv x' \wedge y' \tag{3}$$
$$x \vee y \equiv x' \vee y'. \tag{4}$$

The *equivalence class* of all elements of L equivalent with x is denoted by $[x]$, i.e., $[x] = \{y : y \in L \ \& \ y \equiv x\}$. We use $L \backslash \equiv$ to denote the *quotient lattice* with the equivalence classes $[x]$ as its elements and meet and join operations defined as $[x] \wedge [y] = [x \wedge y]$ and $[x] \vee [y] = [x \vee y]$ for any $[x], [y] \in L \backslash \equiv$. It is easy to see that if $x \equiv y$, then $x \wedge y \equiv x \vee y$. For more information about bounded lattices, we refer to [11, 12, 20].

Definition 1 (*cf.* [17, *Definition 3.1*]) Let $L = (L, \wedge, \vee, 0_L, 1_L)$ be a bounded lattice. An operation $T : L^2 \to L$ ($S : L^2 \to L$) on a bounded lattice L is a *t-norm* (*t-conorm*) if it is commutative, associative, non-decreasing with respect to both variables and 1_L (0_L) is its neutral element, i.e., $T(1_L, x) = x$ ($S(0_L, x) = x$) holds for any $x \in L$.

---

[2]Note that if an index set P is a linearly ordered set of indices with the bottom and top elements, then P is clearly a bounded lattice.

Fig. 1 Illustration of a
lattice-ordered index set P
(left figure) and the
lexicographic product of P
and **2** (right figure)

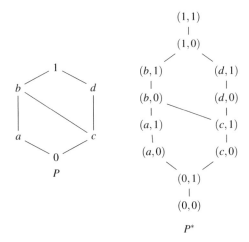

3   Generalized Ordinal Sum of Bounded Lattices

In this section, we introduce the ordinal sum of bounded lattices for an index set that
may not be linearly ordered. More precisely, we assume that each index set has a
structure of a bounded lattice.

Let $P = (P, \wedge, \vee, 0_P, 1_P)$ be an index set, and let $\mathbf{P} = \{L_p : p \in P\}$ be a family
of bounded lattices. We use $\wedge_p$ and $\vee_p$ to denote the meet and join on L_p and 0_p and
1_p the bottom and top elements of L_p, respectively. For the sake of simplicity, we
put $L_p^- = L_p \setminus \{0_p, 1_p\}$. For an ordinal sum construction a simple situation occurs
when the family $\mathbf{P}$ only contains such bounded lattices that are mutually disjoint,
i.e., there is no element that belongs to two different lattices. In this case, Clifford's
idea of the ordinal sum of semigroups can be directly adopted [4].

In general, to get a more complex situation, one can admit a non-empty intersection
of bounded lattices in $\mathbf{P}$. However, the property of non-emptiness must be carefully
treated to design an appropriate general ordinal sum construction. Similarly to [8]
(see also [7]), we assume that $L_p^- \cap L_q^- = \emptyset$ in $\mathbf{P}$ for any $p, q \in P$ such that $p \neq q$.
Then, two different lattices from $\mathbf{P}$ admit the same bottom or top element or both
of these elements or a bottom element of one lattice coincides with a top element
of the second lattice or vice-versa. To model this situation, it seems to be efficient
to consider the bounded lattice P^* instead of the original P, where $(p, 0)$ naturally
serves as an index for 0_p and $(p, 1)$ for 1_p. Particularly, we can define a (set) map

$$\lambda : P^* \to \bigcup_{p \in P} \{0_p, 1_p\} \tag{5}$$

with $\lambda(p, 0) = 0_p$ and $\lambda(p, 1) = 1_p$ for any $p \in P$, and, for example, if $1_p = 1_q$
for $p \neq q$, we have $\lambda(p, 1) = \lambda(q, 1)$. Obviously, the map λ determines an equivalence $\equiv_\lambda$ on P^*, i.e., $(p, i) \equiv_\lambda (q, j)$ whenever $\lambda(p, i) = \lambda(q, j)$, and the question

is, when the quotient set $P^*\backslash \equiv_\lambda$ is well-defined to introduce a generalized ordinal sum of bounded lattices. From the analysis of properties under which the generalized ordinal sum can be correctly introduced, we found that the assumption on $\equiv_\lambda$ to be a congruence on P^* is too strong. Indeed, it is easy to demonstrate that if $p, q \in P$ are such that $p \parallel q$ and simultaneously $(p, 1) \equiv_\lambda (q, 1)$, then $L_p = L_q = \{1_p\}$ which follows from $(p \wedge q, 1) = (p, 1) \wedge (q, 1) \equiv_\lambda (p, 1) \vee (q, 1) = (p, 1) \vee (q, 1) = (p \vee q, 0)$. On the other hand, one can define a generalized ordinal sum of bounded lattices for this case as has been demonstrated in [8]. In the next part, we propose certain properties on $\equiv_\lambda$ ensuring the correctness of our generalized definition of ordinal sum of bounded lattices.

Obviously, if $[x]$ is an equivalence class on a lattice L with respect to a congruence $\equiv$, the class $[x]$ is closed under the meet and join operations, hence, it is a sublattice of L. In the following definition, we consider the closeness only for one of the operations. Recall that each lattice is a *meet (join) semilattice*, where the join (meet) operation is removed, or is not taken into account, respectively. A subset of a meet (join) semilattice L which is itself a meet (join) semilattice with respect to the meet (join) of L is said to be a *meet (join) sub-semilattice*.

Definition 2 Let $\equiv$ be an equivalence on a bounded lattice L. An equivalence class $[x]$ from $L\backslash \equiv$ is said to be *meet (join) closed* if $[x]$ is a meet (join) convex sub-semilattice of the meet (join) semilattice L. A class $[x]$ from $L\backslash \equiv$ is said to be *semi-closed* if it is meet or join closed.

If $L\backslash \equiv$ is a quotient lattice, then $[x] \leq [y]$ if and only if $x' \wedge y' \in [x]$ and $x' \vee y' \in [y]$ hold for any $x' \in [x]$ and $y' \in [y]$. A relaxation of the partial ordering between equivalence classes that are only meet or join closed is as follows.

Theorem 1 *Let $\equiv$ be an equivalence on a bounded lattice L such that $[x]$ is semi-closed for any $x \in L$. Write $[x] \ll [y]$ if $x' \wedge y' \in [x]$ for any $x' \in [x]$ and $y' \in [y]$ or $x' \vee y' \in [y]$ for any $x' \in [x]$ and $y' \in [y]$, and define $[x] \leq [y]$ if there exist $z_1, \ldots, z_n \in L$ such that $[x] \ll [z_1] \ll \cdots \ll [z_n] \ll [y]$. Then $\leq$ is a pre-order on $L\backslash \equiv$.*

Proof Since $[x]$ is meet or join closed for any $x \in L$, we obtain $[x] \ll [x]$, and hence $[x] \leq [x]$. Moreover, if $[x] \leq [y] \leq [z]$, then there exists $z_1, \ldots, z_n, z_1', \ldots, z_m' \in L$ such that

$$[x] \ll [z_1] \ll \cdots \ll [z_n] \ll [y] \ll [z_1'] \ll \cdots [z_m'] \ll [z],$$

hence, we obtain $[x] \leq [z]$. $\qquad \square$

Unfortunately, the previously defined pre-order on $L\backslash \equiv$ is not a partial order in general, because it is not anti-symmetric as can be seen from Fig. 2 where $[g] \leq [c] \leq [g]$, but $[g] \neq [c]$. Note that if the equivalence $\equiv$ is a congruence on L, then the pre-order $\leq$ on $L\backslash \equiv$ becomes a partial order. This fact motivates us to introduce the concept of a semi-congruence on a bounded lattice.

Fig. 2 A pre-ordering
relation on the quotient set
$L\backslash \equiv$ which is not a partial
order

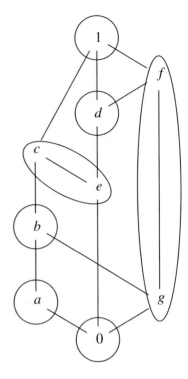

Definition 3 Let L be a bounded lattice. An equivalence $\equiv$ on L is said to be *semi-congruence* if each equivalence class is semi-closed and the pre-order on $L\backslash \equiv$ defined in Theorem 1 is a partial order such that $L\backslash \equiv$ is a lattice.[3]

Let $P = (P, \wedge, \vee, 0_P, 1_P)$ be an index set, and let $p, q \in P$ be two distinct elements. We use $P^*(p, q)$ to denote the least sublattice of the lattice P^* that contains the elements $(p, 0)$, $(p, 1)$, $(q, 0)$ and $(q, 1)$. Let us emphasize that $P^*(p, q)$ need not contain the bottom element 0_{P^*} and the top element 1_{P^*} of P^* because we consider only a sublattice of the lattice P^* and not of the bounded lattice P^*. In Fig. 3, two canonical examples of such sublattices are displayed.

Let $\equiv$ be a semi-congruence on P^*, and let $p, q \in P$ be two distinct elements. Denote $\equiv_{p,q}$ the restriction of $\equiv$ on the sublattice $P^*(p, q)$. It is easy to see that the pre-order on $P^*(p, q)\backslash \equiv_{p,q}$ is always a lattice order where it is sufficient to verify this fact on the canonical sublattices in Fig. 3. Note that the equivalence class $[(p, j)]_{p,q}$ from $P^*(p, q)\backslash \equiv_{p,q}$ is a subset of the equivalence class $[(p, j)]$ in $P^*\backslash \equiv$. Therefore, the natural map $\gamma_{p,q} : P^*(p, q)\backslash \equiv_{p,q} \rightarrow P^*\backslash \equiv$ given by

$$\gamma_{p,q} : ([(p, j)]_{p,q}) = [(p, j)], \quad (p, j) \in P^*(p, q), \tag{6}$$

---

[3] Here we consider the definition of lattice as a partially ordered set such that any two elements have an infimum and a supremum. Note that the algebraic definition of lattice and the definition based on a partially ordered set are equivalent (see, e.g., [12, 20]).

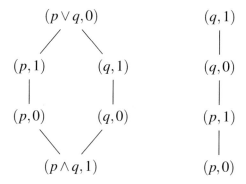

Fig. 3 Two examples of sublattices $P^*(p, q)$ where $p \parallel q$ (left figure) and $p < q$ (right figure)

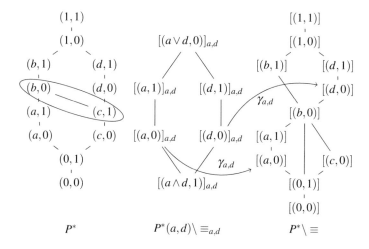

Fig. 4 A semi-congruence on P^* and a map $\gamma_{a,d} : P^*(a, d)\backslash \equiv_{a,d} \rightarrow P^*\backslash \equiv$ which is not an embedding

is well-defined and it is injective, but it need not be a lattice-homomorphism (i.e., an embedding) in general as can be seen in Fig. 4. Here, we consider the lattice-ordered index set P displayed in Fig. 1 and the equivalence class $\equiv$ merging only two different elements $(b, 0)$ and $(c, 1)$. One can see that the equivalence classes $[(a, 0)]$ and $[(d, 0)]$ are incomparable in $P^*(a, d)\backslash \equiv_{a,d}$, but $[(a, 0)] \leq [(d, 0)]$ in $P^*\backslash \equiv$, where the class $[(b, 0)]$ is used to create the ordering relationship.

Since a generalized ordinal sum of bounded lattices must respect, in some sense, the order of elements in P^* for each pair $(p, q) \in P^2$, i.e., if p and q are incomparable in P^*, then it is impossible to alter their order to be comparable in $P^*\backslash \equiv$ where certain elements are merged, we come up with the following definition providing an important condition on a family $\{L_p : p \in P\}$ of bounded lattices.

Definition 4 Let P be a lattice-ordered index set. A family $\{L_p : p \in P\}$ is said to be a *P-family* if $L_p^- \cap L_q^- = \emptyset$ for any $p, q \in P$ such that $p \neq q$ and the map

$$\lambda : P^* \to \bigcup_{p \in P} \{0_p, 1_p\} \tag{7}$$

given by $\lambda(p, 0) = 0_p$ and $\lambda(p, 1) = 1_p$ for any $p \in P$ determines a semi-congruence $\equiv_\lambda$ on P^* such that the natural map $\gamma_{p,q} : P^*(p, q) \backslash \equiv_{\lambda,p,q} \to P^* \backslash \equiv_\lambda$ defined in (6) is a lattice-homomorphism for any $p, q \in P$ with $p \neq q$.

Example 1 Consider the lattice-ordered index set P displayed in Fig. 1 and the bounded lattices in Fig. 4. If $\lambda : P^* \to \bigcup_{p \in P} \{0_p, 1_p\}$ is defined in such a way that only two distinct elements in P^* are merged trough λ, particularly, $\lambda(b, 0) = \lambda(c, 1)$, the family $\mathbf{P} = \{L_p : p \in P\}$ is not a P-family, because $\gamma_{a,d}$ is not a lattice-homomorphism.

The following example shows a family which is a P-family.

Example 2 Let $P = \{0, a, b, c, d, 1\}$ be a bounded lattice displayed in Fig. 1, and let $\mathbf{P} = \{L_p : p \in P\}$ be a family of bounded lattices with the map λ such that $\lambda(p, i) = \lambda(q, j)$ if and only if

$$(p, i), (q, j) \in \{(b, 0), (c, 1)\} \quad \text{or} \quad (p, i), (q, j) \in \{(c, 0), (a, 1), (a, 0), (0, 1)\}.$$

From Fig. 5, one can see that $\equiv_\lambda$ is a semi-congruence on P^*, e.g., the class $[(a, 1)]$ is meet-closed and $[(b, 0)]$ is meet as well as join closed. Consider the lattice $P^*(a, d) \backslash \equiv_{\lambda,a,d}$. In contrast to the previous example, the natural map $\gamma_{a,d}$ is a lattice homomorphism. Similarly, one can verify that $\gamma_{p,q}$ is an embedding for any distinct elements p and q from P. Hence, we obtain that $\mathbf{P}$ is a P-family.

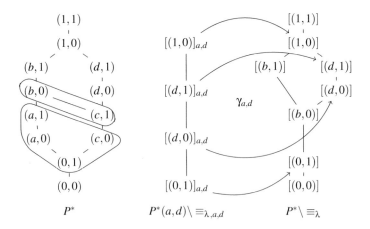

Fig. 5 A semi-congruence on P^* and a map $\gamma_{a,d} : P^*(a, d) \backslash \equiv_{\lambda,a,d} \to P^* \backslash \equiv_\lambda$ which is an embedding

Now, we can introduce the concept of generalized ordinal sum of bounded lattices.

Theorem 2 *Let P be a lattice-ordered index set, and let $\mathbf{P} = \{L_p : p \in P\}$ be a P-family of bounded lattices. Put $L = \bigcup_{p \in P} L_p$ and define the meet and join on L as follows:*

$$
x \wedge y = \begin{cases} x \wedge_p y, & (x, y) \in L_p \times L_p, \\ x, & (x, y) \in L_p \times L_q, \ p < q, \\ y, & (x, y) \in L_p \times L_q, \ q < p, \\ 1_{p \wedge q}, & (x, y) \in L_p \times L_q, \ p \parallel q, \end{cases} \tag{8}
$$

and

$$
x \vee y = \begin{cases} x \vee_p y, & (x, y) \in L_p \times L_p, \\ y, & (x, y) \in L_p \times L_q, \ p < q, \\ x, & (x, y) \in L_p \times L_q, \ q < p, \\ 0_{p \vee q}, & (x, y) \in L_p \times L_q, \ p \parallel q. \end{cases} \tag{9}
$$

Then $L = (L, \wedge, \vee, 0_{0_P}, 1_{1_P})$ is a bounded lattice. The bounded lattice L will be called the P-ordinal sum of bounded lattices and denoted as $L = \bigoplus_{p \in P} L_p$.

4 Interior Operator Based Ordinal Sum of T-Norms

In this section, we extend the results contained in [6]. An important role in our analysis of ordinal sum constructions of t-norms on bounded lattices is played by the concept of an interior operator. Let us recall its definition.

Definition 5 Let L be a bounded lattice. A map $h : L \to L$ is said to be an *interior operator* on L if

1. $h(1_L) = 1_L$,
2. $h(h(x)) = h(x)$ for any $x \in L$,
3. $h(x \wedge y) = h(x) \wedge h(y)$ for any $x, y \in L$,
4. $x \geq h(x)$.

The following two examples show simple constructions of interior operators.

Example 3 Let L be a bounded lattice, and let $b \in L$ be arbitrary. Then, the map $h_b : L \to L$ defined by

$$
h_b(x) = \begin{cases} x, & x \geq b, \\ x \wedge b, & \text{otherwise,} \end{cases} \tag{10}
$$

for any $x \in L$, is an interior operator on L. Note that $h_{0_L} = h_{1_L} = id_L$, where id_L denotes the identity map on L.

Example 4 Let L be a bounded lattice, and let $b \in L \setminus \{0_L, 1_L\}$ be arbitrary. Then, the map $h_b : L \to L$ defined by

$$h_b(x) = \begin{cases} x, & x \geq b, \\ 0_L, & \text{otherwise,} \end{cases} \tag{11}$$

for any $x \in L$, is an interior operator on L.

One can simply verify that the composition of interior operators on a bounded lattice L from the previous two examples is commutative, more precisely, $h_a \circ h_b = h_b \circ h_a$ holds for any $a, b \in L$ with $a \leq b$. And moreover, their composition is again an interior operator. This motivates us to introduce the following definition.

Definition 6 Let $h, g : L \to L$ be maps on a bounded lattice L. We say that h and g commute on L provided that $h \circ g = g \circ h$.

The commutativity of interior operators is a sufficient condition to ensure that their composition is again an interior operator as the following lemma states.

Lemma 1 Let h, g be interior operators on a bounded lattice L that commute on L. Then $h \circ g$ is an interior operator on L.

The following theorem is crucial for a construction of t-norms on bounded lattices with the help of an interior operator. The proof of this theorem can be found in [6].

Theorem 3 Let L be a bounded lattice, and let $h : L \to L$ be an interior operator on L. Let H denote the image of L under h, i.e., $h(L) = H$. If V is a t-norm on H, then there exists its extension to a t-norm T on L as follows:

$$T(x, y) = \begin{cases} V(h(x), h(y)), & x, y \in L \setminus \{1_L\}; \\ x \wedge y, & \text{otherwise.} \end{cases}$$

The next theorem introduces the ordinal sum of t-norms on a bounded lattice, which is an ordinal sum of bounded lattices over a P-family $\mathbf{P} = \{L_p : p \in P\}$, where P is a bounded lattice.

Theorem 4 Let $\mathbf{P} = \{L_p : p \in P\}$ be a P-family, and let $L = \bigoplus_{p \in P} L_p$. If V_p is a t-norm on L_p for any $p \in P$, then the ordinal sum of t-norms $\{V_p \mid p \in P\}$ defined as follows:

$$T(x, y) = \begin{cases} V_p(x, y), & x, y \in L_p \setminus \{1_p\}, \\ x \wedge y, & \text{otherwise,} \end{cases} \tag{12}$$

where $\wedge$ is defined by (8), is a t-norm on L.

Theorem 5 *Let L be a bounded lattice, let $P = \{L_p : p \in P\}$ be a P-family, and let $K = \bigoplus_{p \in P} L_p$ such that K is a bounded sublattice of L. If V_p is a t-norm on L_p for any $p \in P$ and h is an interior operator on L such that $h(L) = K$, then*

$$T(x, y) = \begin{cases} V_p(h(x), h(y)), & (h(x), h(y)) \in L_p \setminus \{1_p\}^2, \\ h(x) \wedge h(y), & (h(x), h(y)) \in L_p \setminus \{1_p\} \times L_q \setminus \{1_q\}, \text{ for } p \neq q, \\ x \wedge y, & \text{otherwise,} \end{cases}$$

(13)

for any $x, y \in L$, is a t-norm on L, which is called the h-ordinal sum of t-norms $\{V_p \mid p \in P\}$.

Example 5 Let L be a bounded lattice, and let P be a lattice-ordered index set such that there is a maximal finite chain $0_P = p_0 < p_1, \ldots < p_n = 1_P$. Assume that $\mathbf{P} = \{L_p : p \in P\}$ is a P-system of bounded lattices, and denote K their ordinal sum, which is a bounded sublattice of L. Put $M = \bigcup_{i=0}^{n} L_{p_i}$ and $N = K \setminus M$. Consider the interior operator $h_{1_{p_i}} : L \to L$ defined by (cf. (10) in Ex. 3):

$$h_{1_{p_i}}(x) = \begin{cases} x, & x \in N \cup ((1_{p_i}, 1_L] \cap K), \\ x \wedge 1_{p_i}, & \text{otherwise,} \end{cases}$$

(14)

for any $i = 1, \ldots, n - 1$ and construct the operator h as the composition of $h_{1_{p_i}}, i = 1, \ldots, n - 1$. Since these interior operators are mutually commutative, by Lemma 1 we obtain that h is an interior operator on L such that $h(L) = K$. Note that if $h_{p_i}(x) > 0_{p_i}$ for a certain $x \in L \setminus K$ but $h_{p_i}(x) \notin L_{p_i}$ (L_{p_i} need not be an interval in L), then $h(x) = 1_{p_{i-1}} \in L_{p_{i-1}}$. Using (13), one can introduce a t-norm on L.

It should be noted that the presented results can be dually formulated for t-conorms. Particularly, if one replaces the t-norm operation, the meet, the top element and the interior operator in Theorems 3–5 by the t-conorm operation, the join, the bottom element and the closure operator,[4] respectively, the dual statements for the t-conorms can be obtained.

5 Conclusion

In this paper, we analyzed conditions under which an ordinal sum of bounded lattices can be introduced from a family of bounded lattices. We proposed a concept of a semi-congruence and stated a condition ensuring that a relationship between two indexes is preserved when the certain bottom and top elements coincide in the family of

---

[4]The closure operator is defined dually to the interior operator, i.e., $h : L \to L$ is a *closure operator* if 1. $h(0_L) = 0_L$, 2. $h(h(x)) = h(x)$, 3. $h(x \vee y) = h(x) \vee h(y)$, 4. $x \leq h(x)$ holds for any $x, y \in L$. For details, we refer to [6].

bounded lattices. Based on this construction we proposed an extension of an ordinal sum construction of t-norms on bounded lattices [6], called h-ordinal sum, which generalized previous ordinal sum approaches using the concept of lattice interior operator. In further research, we plan to study similar constructions for other classes of aggregation operators.

References

1. Baets, B.D., Mesiar, R.: Triangular norms on product lattices. Fuzzy Sets Syst. **104**, 61–75 (1999)
2. Birkhoff, G.: Lattice Theory. American Mathematical Society, Providence (1973)
3. Çayli, G.D.: On a new class of t-norms and t-conorms on bounded lattices. Fuzzy Sets Syst. **332**, 129–143 (2018)
4. Clifford, A.H.: Naturally totally ordered commutative semigroups. Am. J. Math. **76**, 631–646 (1954)
5. de Cooman, G., Kerre, E.E.: Order norms on bounded partially ordered sets. J. Fuzzy Math. **2**, 281–310 (1994)
6. Dvořák, A., Holčapek, M.: New construction of an ordinal sum of t-norms and t-conorms on bounded lattices. Inf. Sci. **515**, 116–131 (2020)
7. El-Zekey, M.: Lattice-based sum of t-norms on bounded lattices. Fuzzy Sets Syst. (2019). https://doi.org/10.1016/j.fss.2019.01.006
8. El-Zekey, M., Medina, J., Mesiar, R.: Lattice-based sums. Inf. Sci. **223**, 270–284 (2013)
9. Ertuğrul, U., Karaçal, F., Mesiar, R.: Modified ordinal sums of triangular norms and triangular conorms on bounded lattices. Int. J. Intell. Syst. **30**, 807–817 (2015)
10. Goguen, J.A.: L-fuzzy sets. J. Math. Anal. Appl. **18**, 145–174 (1967)
11. Grätzer, G.: General Lattice Theory. Academic Press, New York (1978)
12. Grätzer, G.: Lattice Theory: Foundation. Birkhäuser, Basel (2011)
13. Klement, E.P., Mesiar, R., Pap, E.: Triangular Norms. Trends in Logic, vol. 8. Kluwer, Dordrecht (2000)
14. Medina, J.: Characterizing when an ordinal sum of t-norms is a t-norm on bounded lattices. Fuzzy Sets Syst. **202**, 75–88 (2012)
15. Mostert, P.S., Shields, A.L.: On the structure of semigroups on a compact manifold with boundary. Ann. Math. Second Ser. **65**, 117–143 (1957)
16. Rutherford, D.E.: Introduction to Lattice Theory. Oliver & Boyd, Edinburgh and London (1965)
17. Saminger, S.: On ordinal sums of triangular norms on bounded lattices. Fuzzy Sets Syst. **157**, 1403–1416 (2006)
18. Saminger-Platz, S., Klement, E.P., Mesiar, R.: On extensions of triangular norms on bounded lattices. Indag. Math.-New Ser. **19**, 135–150 (2008)
19. Schweizer, B., Sklar, A.: Probabilistic Metric Spaces. North-Holland, New York (1983)
20. Szász, G.: Introduction to Lattice Theory. Academic Press, New York (1963)
21. Zhang, D.: Triangular norms on partially ordered sets. Fuzzy Sets Syst. **153**, 195–209 (2005)

A Comparison of Some t-Norms and t-Conorms over the Steady State of a Fuzzy Markov Chain

Juan Carlos Figueroa-García

Abstract This chapter shows a comparison of several t-norms and t-conorms used to compute the steady state of a fuzzy Markov chain. The effect of every of the selected norms over the mean and variance of a fuzzy Markov chain is evaluated using some simulations. Some recommendations and concluding remarks are given.

Keywords Fuzzy t-norms · Fuzzy t-conorms · Fuzzy Markov chains

1 Introduction

Fuzzy Markov chains is one of the most popular fuzzy stochastic process (see Sanchez [16, 17], Avrachenkov and Sanchez [2, 3], and Araiza et al. [1]) so there are different algorithms, fuzzy relations and compositions to compute the stable state possibilities of the fuzzy process.

Most of published works on fuzzy Markov chains apply the max-min composition to obtain the steady state of the process (see Figueroa-García [7, 8]), so there is the possibility of using other compositions to see the effect over its steady state.

It is well known that different t-norms and t-conorms lead to closer/wider results in fuzzy inference systems, so we want to evaluate the effect of several norms such as Gödel, Archimedean, Łukasiewicz, Hamacher, Sugeno-Weber, and Yager t-norms (see Klement et al. [13]) over the mean/variance of a fuzzy Markov chain (a.k.a weak inference of a stochastic process).

J. C. Figueroa-García (✉)
Universidad Distrital Francisco José de Caldas, Bogotá, Colombia
e-mail: jcfigueroag@udistrital.edu.co

© Springer Nature Switzerland AG 2020
M. Ceberio and V. Kreinovich (eds.), *Decision Making under Constraints*,
Studies in Systems, Decision and Control 276,
https://doi.org/10.1007/978-3-030-40814-5_10

2 Fuzzy Markov Chains

Let $\mathscr{P}(X)$ be the class of all crisp sets, $\mathscr{F}(X)$ is the class of all fuzzy sets, $\mathscr{F}_1(X)$ is the class of all convex fuzzy sets, and $I = [0, 1]$ be the set of values in the unit interval. A fuzzy set namely A is characterized by a membership function $\mu_A : X \rightarrow I$ defined over a universe of discourse $x \in X$. Thus, a fuzzy set A is the set of ordered pairs $x \in X$ and its membership degree, $\mu_A(x)$, i.e.,

$$A = \{(x, \mu_A(x)) \mid x \in X\}. \tag{1}$$

Avrachenkov and Sanchez [3] defined a fuzzy Markov chain in a similar way than a probabilistic Markov chain, as shown as follows.

Definition 1 Let $S = \{1, 2, \ldots, n\}$, then a finite fuzzy set or a fuzzy distribution on S is defined by a map $x : S \rightarrow [0, 1]$ represented by a vector $\mathbf{x} = \{x_1, x_2, \ldots, x_n\}$ where $0 \leqslant x_i \leqslant 1, i \in S$. The set of all fuzzy sets is denoted by $\mathscr{F}(S)$.

A fuzzy distribution for a $S = 33$ states is shown in Fig. 1.

A fuzzy relation matrix P on the cartesian product $S \times S$ where P is defined by a matrix $\{p_{ij}\}_{i,j=1}^n$, with $0 \leqslant p_{ij} \leqslant 1, i, j \in S$ i.e.

$$P = \begin{bmatrix} p_{11} & p_{12} & \cdots & p_{1m} \\ p_{21} & p_{22} & \cdots & p_{2m} \\ \vdots & \vdots & \ddots & \vdots \\ p_{m1} & p_{m2} & \cdots & p_{mm} \end{bmatrix}$$

This fuzzy relationship matrix P allows us to define the transition matrix of the m states of the Markov chain between each time instant t in the following form:

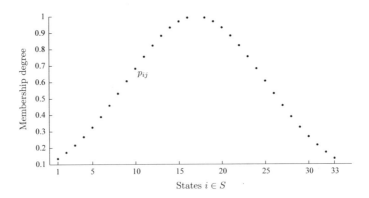

Fig. 1 Fuzzy distribution for $S = 33$ states

Definition 2 At each instant t, $t = 1, 2, \ldots, n$, the state of the stochastic process is described by the fuzzy set $x^{(t)} \in \mathscr{F}(S)$. The transition law of the Markov chain is given by the fuzzy relation P as follows, at the instant t, $T = 1, 2, \ldots, n$.

$$\mathbf{x}^{(t+1)} = \bigvee_{i \in S} \{\mathbf{x}^{(t)} \wedge p_{ij}\}, \ j \in S. \tag{2}$$

were i and j are the initial and final state of the transition i, $j = 1, 2, \ldots, m$ and $\mathbf{x}^{(0)}$ is the initial fuzzy set or the initial distribution of the process.

Like in probabilistic Markov chains, the time-limit behavior of a fuzzy transition matrix can be found using the powers of a fuzzy transition matrix P i.e.

$$p_{ij}^t \overset{\triangle}{=} \bigvee_{k \in S} \{p_{ik} \wedge p_{kj}^{t-1}\} \tag{3}$$

or in a matrix form:

$$P^t \overset{\triangle}{=} P \circ P^{t-1} \tag{4}$$

where $\circ$ denotes a fuzzy composition, and the limiting distribution of P is:

Definition 3 Let the powers of the fuzzy transition matrix converge in τ steps to a non periodic solution, then the associated fuzzy Markov chain is called non periodic (or aperiodic) and $P^* = P^\tau$ is called a limiting fuzzy transition matrix i.e.

$$\lim_{n \to \tau} P^n = P^*.$$

In the **probabilistic approach**, the powers of P are computed using the sum-prod composition (Chapman-Kolmogorov equation) as follows:

$$p_{ij}^t \overset{\triangle}{=} \bigvee_{k \in S} \{p_{ik} \wedge p_{kj}^{t-1}\} = \sum_{k \in S} p_{ik} \cdot p_{kj}^{t-1} \tag{5}$$

$$\sum_{j \in S} p_{ij} = 1 \ \forall i \in S \tag{6}$$

which is known as the convexity law of a probability distribution.

2.1 Ergodicity in FMs

While most of probabilistic Markov chains exhibit some steady state (even periodical), the idea of ergodicity in FMs is not as exact as in probabilistic Markov chains since it depends on the selected t-norms and t-conorms. Figueroa-García [7, 8] defined three basic concepts around ergodicity for FMs.

Definition 4 (*Strong ergodicity*) A fuzzy transition matrix P is said to be **strong ergodic** if its powers $\lim_{n \to \tau} P^n = P^*$ converge to an idempotent matrix with equal rows/columns.

Definition 5 (*Weak ergodicity*) A fuzzy transition matrix P is said to be **weak ergodic** if its powers $\lim_{n \to \tau} P^n = P^*$ converge to an idempotent matrix with different rows/columns.

Definition 6 (*Periodicity*) A fuzzy transition matrix P is said to be **periodic** if its powers $\lim_{n \to \tau} P^n = P^*$ into an oscillating matrix with period k.

Martin Gavalec (see [10–12]) pointed out the NP-hard nature of computing the powers of a fuzzy matrix. Based on his results, Avrachenkov and Sanchez (see [2, 3]) proposed three algorithms to compute the final stationary possibility distribution of P namely $\mathbf{p}^*$:

$$\mathbf{p}^* = \max_{i \in S}\{p_{ij}^*\}, \ j \in S. \tag{7}$$

Despite the problem of computing P^n is very interesting, the focus of this chapter is to see how different t-norms and t-conorms behave in a time-limit criterion.

2.2 Selected t-Norms and t-Conorms

In this chapter, we have selected some t-norms and t-conorms (logics) to compute the transition law of an FM (see Eq. (2)):

$$\text{max-min} \overset{\triangle}{=} \max_{k \in S} \ \min\{p_{ik}, p_{kj}^{t-1}\} \tag{8}$$

$$\text{max-prod} \overset{\triangle}{=} \max_{k \in S}\{p_{ik} \cdot p_{kj}^{t-1}\} \tag{9}$$

$$\text{max-Łuk.} \overset{\triangle}{=} \max_{k \in S}\{p_{ik} + p_{kj}^{t-1} - 1, 0\} \tag{10}$$

$$\text{max-Ham.} \overset{\triangle}{=} \max_{k \in S}\left\{\frac{p_{ik} \cdot p_{kj}^{t-1}}{p_{ik} + p_{kj}^{t-1} - p_{ik} \cdot p_{kj}^{t-1}}\right\} \tag{11}$$

$$\text{max-Yager} \overset{\triangle}{=} \max_{k \in S} \ \max\{0, 1 - ((1 - p_{ik})^a + (1 - p_{kj}^{t-1})^a)^{1/a}\}, \ a \in \mathbb{N} \tag{12}$$

Every logic has its own behavior, some of them are wider than others which are more conservative. It is well known that the max-min logic is one of the most conservative, and the max-Hamacher logic gets much more unstable results, so they affect the steady state of a fuzzy Markov chain in different ways.

3 An Example

As pointed out by Figueroa-García et al. (see [7, 8]) the max-min logic leads to either weak ergodic or periodic behaviors, so the following example is intended to show the behavior of the steady state of a fuzzy transition matrix computed using the t-norms and t-conorms shown in the previous section. Let us define the following transition matrix:

$$P = \begin{bmatrix} 0.441 & 0.957 & 0.992 & 0.680 & 0.033 \\ 0.550 & 0.384 & 0.042 & 0.186 & 0.913 \\ 0.296 & 0.520 & 0.999 & 0.305 & 0.423 \\ 0.002 & 0.634 & 0.340 & 0.051 & 0.964 \\ 0.689 & 0.680 & 0.524 & 0.848 & 0.278 \end{bmatrix}$$

Note that P is non-normal, so we first normalize $\hat{p}_{ij}$ using the max method:

$$\hat{p}_{ij} = \frac{p_{ij}}{\max_j\{p_{ij}\}} \ \forall \ i \in S$$

which results into the following normalized matrix $\hat{P}$:

$$\hat{P} = \begin{bmatrix} 0.445 & 0.964 & 1 & 0.685 & 0.034 \\ 0.603 & 0.421 & 0.046 & 0.203 & 1 \\ 0.296 & 0.520 & 1 & 0.305 & 0.424 \\ 0.002 & 0.658 & 0.353 & 0.053 & 1 \\ 0.813 & 0.803 & 0.618 & 1 & 0.328 \end{bmatrix}$$

To evaluate the effect of the selected logics over the behavior of the stochastic process, we compute the expected value and variance of $\mathbf{p}^*$, as defined as follows:

$$E(P) = \frac{\sum_j j \cdot p_j^*}{\sum_j p_j^*}, \tag{13}$$

$$V(P) = \frac{\sum_j (j - E(P))^2 \cdot p_j^*}{\sum_j p_j^*} \tag{14}$$

where p_j^* is the j_{th} element of the limiting distribution of $\mathbf{p}^*$ (see Eq. (7)).

3.1 Stationary Transition Matrices

After computing an amount τ of powers of $\hat{P}$, it gets either a weak/strong ergodic or periodic behavior. The steady state $\hat{P}^*$ for every logic is presented next.

max-min logic

The stationary matrix $\hat{P}^*$ is obtained in $\tau = 16$ powers.

$$\hat{P}^* = \begin{bmatrix} 0.813 & 0.813 & 1 & 0.813 & 0.964 \\ 0.813 & 0.813 & 0.813 & 1 & 0.813 \\ 0.520 & 0.520 & 1 & 0.520 & 0.520 \\ 0.813 & 0.813 & 0.813 & 1 & 0.813 \\ 0.813 & 0.813 & 0.813 & 0.813 & 1 \end{bmatrix}$$

The values of $\hat{\mathbf{p}}^*$, $E(P)$ and $V(P)$ are shown next.

$$\hat{\mathbf{p}}^* = [0.813\ 0.813\ 1\ 1\ 1]$$
$$E(P) = 3.1215$$
$$V(P) = 1.9448$$

max-prod logic

The stationary matrix $\hat{P}^*$ is obtained in $\tau = 16$ powers.

$$\hat{P}^* = \begin{bmatrix} 0.614 & 0.756 & 1 & 0.756 & 0.964 \\ 0.813 & 0.803 & 0.813 & 1 & 0.784 \\ 0.423 & 0.520 & 1 & 0.520 & 0.520 \\ 0.813 & 0.803 & 0.813 & 1 & 0.784 \\ 0.637 & 0.784 & 0.813 & 0.784 & 1 \end{bmatrix}$$

The values of $\hat{\mathbf{p}}^*$, $E(P)$ and $V(P)$ are shown next.

$$\hat{\mathbf{p}}^* = [0.813\ 0.803\ 1\ 1\ 1]$$
$$E(P) = 3.1239$$
$$V(P) = 1.9463$$

max-Łukasiewicz logic

The stationary matrix $\hat{P}^*$ is obtained in $\tau = 16$ powers.

$$\hat{P}^* = \begin{bmatrix} 0.554 & 0.741 & 1 & 0.741 & 0.964 \\ 0.813 & 0.803 & 0.813 & 1 & 0.777 \\ 0.333 & 0.520 & 1 & 0.520 & 0.520 \\ 0.813 & 0.803 & 0.813 & 1 & 0.777 \\ 0.590 & 0.777 & 0.813 & 0.777 & 1 \end{bmatrix}$$

The values of $\hat{\mathbf{p}}^*$, $E(P)$ and $V(P)$ are shown next.

$$\hat{\mathbf{p}}^* = [0.813\ 0.803\ 1\ 1\ 1]$$
$$E(P) = 3.1239$$
$$V(P) = 1.9463$$

max-Hamacher logic

The stationary matrix $\hat{P}^*$ is obtained in $\tau = 1024$ powers and it exhibits a period of 16.

$$\hat{P}^* = \begin{bmatrix} 0.651 & 0.779 & 1 & 0.964 & 0.766 \\ 0.813 & 0.789 & 0.813 & 0.789 & 1 \\ 0.465 & 0.520 & 1 & 0.520 & 0.520 \\ 0.813 & 0.789 & 0.813 & 0.789 & 1 \\ 0.668 & 0.803 & 0.813 & 1 & 0.789 \end{bmatrix}$$

The values of $\hat{\mathbf{p}}^*$, $E(P)$ and $V(P)$ are shown next.

$$\hat{\mathbf{p}}^* = [0.813\ 0.803\ 1\ 1\ 1]$$
$$E(P) = 3.1239$$
$$V(P) = 1.9463$$

max-Yager logic

The stationary matrix $\hat{P}^*$ is obtained in $\tau = 16$ powers.

$$\hat{P}^* = \begin{bmatrix} 0.777 & 0.813 & 1 & 0.813 & 0.964 \\ 0.813 & 0.803 & 0.813 & 1 & 0.813 \\ 0.518 & 0.520 & 1 & 0.520 & 0.520 \\ 0.813 & 0.803 & 0.813 & 1 & 0.813 \\ 0.777 & 0.813 & 0.813 & 0.813 & 1 \end{bmatrix}$$

The values of $\hat{\mathbf{p}}^*$, $E(P)$ and $V(P)$ are shown next.

$$\hat{\mathbf{p}}^* = [0.813\ 0.803\ 1\ 1\ 1]$$
$$E(P) = 3.1239$$
$$V(P) = 1.9463$$

3.2 Scalarization of P

Another way to obtain the steady state of a fuzzy Markov chain is scaling P using the method proposed by Figueroa-García et al. [8]. It is basically a method in which every vector $i \in S$ is normalized using the total sum of their memberships, as shown as follows:

$$\hat{p}_{ij} = \frac{p_{ij}}{\sum_j p_{ij}}, \ (i, j) \in S$$

An interesting property of this scalarization method is that it obtains a convex matrix P i.e.

$$\sum_j \hat{p}_{ij} = 1, \ \forall i \in S$$

which is equivalent to have a probabilistic Markov chain. The obtained results over the application example are:

$$\hat{P} = \begin{bmatrix} 0.142 & 0.308 & 0.320 & 0.219 & 0.011 \\ 0.265 & 0.185 & 0.020 & 0.089 & 0.440 \\ 0.116 & 0.204 & 0.393 & 0.120 & 0.166 \\ 0.001 & 0.319 & 0.171 & 0.025 & 0.484 \\ 0.228 & 0.225 & 0.174 & 0.281 & 0.092 \end{bmatrix}$$

The stationary matrix $\hat{P}^*$ is obtained in $\tau = 16$ powers.

$$\hat{P}^* = \begin{bmatrix} 0.165 & 0.239 & 0.206 & 0.153 & 0.237 \\ 0.165 & 0.239 & 0.206 & 0.153 & 0.237 \\ 0.165 & 0.239 & 0.206 & 0.153 & 0.237 \\ 0.165 & 0.239 & 0.206 & 0.153 & 0.237 \\ 0.165 & 0.239 & 0.206 & 0.153 & 0.237 \end{bmatrix}$$

The values of $\hat{\mathbf{p}}^*$, $E(P)$ and $V(P)$ are shown next.

$$\hat{\mathbf{p}}^* = [0.165 \ 0.239 \ 0.206 \ 0.153 \ 0.237]$$
$$E(P) = 3.0570$$
$$V(P) = 1.9974$$

3.3 Analysis of the Results

The first thing we found in this example is that all fuzzy approaches led to weak strong stationary possibility distributions while the scalarization approach obtained a strong stationary distribution. It is interesting to see that all $\hat{P}^*$ are very different, but keep a unitary value on each vector (it is an evidence of dominance).

Now, all the compositions (except the max-min composition) obtained the same stationary possibility distribution $\mathbf{p}^*$. This is an evidence that (in the limit) there is a stationary distribution where all compositions are converging to. This leads us to think that there is a possibility of having dominance between memberships.

It is important to note that $E(P)$ and $V(P)$ are very similar in all approaches even while the scalarized method obtained a slightly smaller $E(P)$ and bigger $V(P)$. This could mean that the example is having a expected steady state, and the fuzzy composition used to obtain its stationary distribution is only a way to head to its time-limit behavior.

4 Concluding Remarks

Six different compositions to compute the powers of a fuzzy matrix P were applied to an example and obtained similar stationary distributions for P^*. All pure fuzzy compositions obtained weak ergodic distributions while a scalarization method obtained an ergodic fuzzy-probabilistic distribution.

There is evidence of dominance among memberships (at least on its stationary distribution) since all fuzzy compositions led to very similar stationary distributions. The fuzzy-probabilistic approach does not show this dominance due to the continuous nature of the *sum-prod* composition plus the convexity of the normalized transition matrix.

Finally, it is important to remark that most of fuzzy Markov chains do not converge to a strong ergodic distribution as their probabilistic counterparts (see Figueroa-García et al. [7, 8]) so the concept of ergodicity and convergence in fuzzy stochastic processes needs additional definitions and methods.

Future works on fuzzy Markov chains include interval-valued Markov chains (see Skulj [18], Andrade-Campos [4], Kurano et al. [14, 15]), Type-2 fuzzy Markov chains (see Zeng and Liu [19], Figueroa-García [5, 6, 9]) and fuzzy queueing theory.

Acknowledgements Juan Carlos Figueroa-Garcia would like to thank to his mother Maria Irene García for being the most important person in his life.

References

1. Araiza, R., Xiang, G., Kosheleva, O., Skülj, D.: Under interval and fuzzy uncertainty, symmetric Markov chains are more difficult to predict. In: 2007 Annual Meeting of the North American Fuzzy Information Processing Society (NAFIPS), vol. 26, pp. 526–531. IEEE (2007)
2. Avrachenkov, K.E., Sanchez, E.: Fuzzy Markov chains: specificities and properties. In: IEEE (ed.) 8th IPMU'2000 Conference, Madrid, Spain, pp. 1851–1856. IEEE (2000)
3. Avrachenkov, K.E., Sanchez, E.: Fuzzy Markov chains and decision-making. Fuzzy Optim. Decis. Mak. **1**(2), 143–159 (2002)
4. Campos, M.A., Dimuro, G.P., da Rocha Costa, A.C., Kreinovich, V.: Computing 2-step predictions for interval-valued finite stationary Markov chains. Tech. rep., UTEP-CS-03-20 (2003)
5. Figueroa-García, J.C.: Interval type-2 fuzzy Markov chains: an approach. In: 2007 Annual Meeting of the North American Fuzzy Information Processing Society (NAFIPS), vol. 28, No. 1, pp. 1–6 (2010)
6. Figueroa-García, J.C.: Advances in Type-2 Fuzzy Sets and Systems, vol. 301, chap. Interval Type-2 Fuzzy Markov Chains, pp. 49–64. Springer (2013)

7. Figueroa-García, J.C., Kalenatic, D., Lopéz, C.A.: A simulation study on fuzzy Markov chains. Commun. Comput. Inf. Sci. **15**(1), 109–117 (2008)
8. Figueroa-García, J.C., Kalenatic, D., Lopez, C.A.: Scalarization of fuzzy Markov chains. Lect. Notes Comput. Sci. **6215**(1), 110–117 (2010)
9. Figueroa-García, J.C., Kalenatic, D., López-Bello, C.A.: Interval type-2 fuzzy Markov chains: type reduction. Lect. Notes Comput. Sci. **6839**(1), 211–218 (2011)
10. Gavalec, M.: Reaching matrix period is np-complete. Tatra Mt. Math. Publ. **12**(1), 81–88 (1997)
11. Gavalec, M.: Periods of special fuzzy matrices. Tatra Mt. Math. Publ. **16**(1), 47–60 (1999)
12. Gavalec, M.: Computing orbit period in max-min algebra. Discret. Appl. Math. **100**(1), 49–65 (2000)
13. Klement, E.P., Mesiar, R., Pap, E.: Triangular Norms. Kluwer Academic Publishers (2009)
14. Kurano, M., Yasuda, M., Nakagami, J.: Interval methods for uncertain Markov decision processes. In: International Workshop on Markov Processes and Controlled Markov Chains (1999)
15. Kurano, M., Yasuda, M., Nakagami, J., Yoshida, Y.: A fuzzy approach to Markov decision processes with uncertain transition probabilities. Fuzzy Sets Syst. **157**(16), 2674–2682 (2007)
16. Sanchez, E.: Resolution of eigen fuzzy sets equations. Fuzzy Sets Syst. **1**(1), 69–74 (1978)
17. Sanchez, E.: Eigen fuzzy sets and fuzzy relations. J. Math. Anal. Appl. **81**(1), 399–421 (1981)
18. Skulj, D.: Regular finite Markov chains with interval probabilities. In: 5th International Symposium on Imprecise Probability: Theories and Applications, Prague, Czech Republic, 2007 (2007)
19. Zeng, J., Liu, Z.Q.: Type-2 fuzzy Markov random fields to handwritten character recognition. In: 18th International Conference on Proceedings of Pattern Recognition, 2006, ICPR 2006, vol. 1, pp. 1162–1165 (2006)

Plans Are Worthless but Planning Is Everything: A Theoretical Explanation of Eisenhower's Observation

Angel F. Garcia Contreras, Martine Ceberio and Vladik Kreinovich

Abstract The 1953–1961 US President Dwight D. Eisenhower emphasized that his experience as the Supreme Commander of the Allied Expeditionary Forces in Europe during the Second World War taught him that "plans are worthless, but planning is everything". This sound contradictory: if plans are worthless, why bother with planning at all? In this paper, we show that Eisenhower's observation has a meaning: while directly following the original plan in constantly changing circumstances is often not a good idea, the existence of a pre-computed original plan enables us to produce an almost-optimal strategy—a strategy that would have been computationally difficult to produce on a short notice without the pre-existing plan.

1 Introduction: Eisenhower's Seemingly Paradoxical Observation

Eisenhower's observation. Dwight D. Eisenhower, the Supreme Commander of the Allied Expeditionary Forces in Europe during the Second World War and later the US President, emphasized that his war experience taught him that "plans are worthless, but planning is everything"; see, e.g. [1].

At first glance, this observation seems paradoxical. At first glance, the Eisenhower's observation sounds paradoxical: if plans are worthless, why bother with planning at all?

What we do in this paper. In this paper, we show that this Eisenhower's observation has a meaning. Namely, it means that:

A. F. Garcia Contreras · M. Ceberio · V. Kreinovich (✉)
University of Texas at El Paso, El Paso, TX 79968, USA
e-mail: vladik@utep.edu

A. F. Garcia Contreras
e-mail: afgarciacontreras@miners.utep.edu

M. Ceberio
e-mail: mceberio@utep.edu

© Springer Nature Switzerland AG 2020
M. Ceberio and V. Kreinovich (eds.), *Decision Making under Constraints*,
Studies in Systems, Decision and Control 276,
https://doi.org/10.1007/978-3-030-40814-5_11

- while following the original plan in constantly changing circumstances is often not a good idea,
- the existence of a pre-computed original plan enables us to produce an almost-optimal strategy (a strategy that would have been computationally difficult to produce on a short notice without the pre-existing plan).

2   Analysis of the Problem

Rational decision making: a brief reminder. According to decision making theory, decisions by a rational decision maker can be described as maximizing the value a certain function known as utility; see, e.g. [3, 4]. In financial situations, when a company needs to make a decision, the overall profit can be used as the utility value; in more complex situations, the utility function combines different aspects of gain and loss related to different decisions.

Let us describe this in precise terms. Let x denote a possible action, a describes the situation, and let $u(x, a)$ denote the utility that results from performing action x in situation a.

To describe a possible action, we usually need to describe the values of several different quantities. For example, a decision about a plant involves selecting amount of gadgets of different type manufactured at this plant—and maybe also the parameters characterizing these gadgets. Let us denote the parameters describing an action by $x_1, \ldots, x_n$. In these terms, an action can be characterized by the tuple

$$x = (x_1, \ldots, x_n).$$

Similarly, in general, we need several different quantities to describe a situation, so we will describe a situation by a tuple $a = (a_1, \ldots, a_m)$.

In these terms, what is planning. Let $\tilde{a}$ describe the original situation. Based on this situation, we come up with an action $\tilde{x}$ that maximizes the corresponding utility: $u(\tilde{x}, \tilde{a}) = \max_x u(x, \tilde{a})$. Computing this optimal action $\tilde{x}$ is what we usually call *planning*.

Situations change. At the moment when we need to start acting, the situation may have changed in comparison with the original situation $\tilde{a}$, to a somewhat different situation a. Let us denote the corresponding change by $\Delta a \overset{\text{def}}{=} a - \tilde{a}$. In terms of this difference, the new situation takes the form $a = \tilde{a} + \Delta$.

A not-always-very-good option: applying the original plan to the new situation. One possibility is to simply ignore the change, and apply the original plan $\tilde{x}$—which was optimal for the original situation $\tilde{a}$—to the new situation $a = \tilde{a} + \Delta a$.

This plan is, in general, not optimal for the new situation. Thus, in comparison to the actually optimal plan x^{opt} for which

$$u(x^{\text{opt}}, \tilde{a} + \Delta a) = \max_x u(x, \tilde{a} + \Delta a),$$

we lose the amount $L_0 \stackrel{\text{def}}{=} u(x^{\text{opt}}, \tilde{a} + \Delta a) - u(\tilde{x}, \tilde{a} + \Delta a)$.

A better option: trying to modify the original plan. Why cannot we just find the optimal solution for the new situation? Because optimization is, in general, an NP-hard problem (see, e.g. [2, 5]), meaning that it is not possible to find the exact optimum in reasonable time.

What we can do is try to use some feasible algorithm—e.g., solving a system of linear equations—to replace the original plan $\tilde{x}$ with a modified plan $\tilde{x} + \Delta x$. Due to NP-hardness, this feasibly modified plan is, in general, not optimal, but we hope that the resulting loss $L_1 \stackrel{\text{def}}{=} u(x^{\text{opt}}, \tilde{a} + \Delta a) - u(\tilde{x} + \Delta x, \tilde{a} + \Delta a)$ is much smaller than the loss L_0 corresponding to the use of the original plan $\tilde{x}$.

What we do in this paper. In this paper, we analyze the values of both losses and we show that indeed, L_1 is much smaller than L_0. So, in many situations, even if the loss L_0 is so large that the corresponding strategy (of directly using the original plan) is worthless, the modified plan may leads to a reasonably small loss $L_1 \ll L_0$—thus explaining Eisenhower's observation.

Estimating L_0. We assume that the difference Δa is reasonably small, so the corresponding difference in action $\Delta x^{\text{opt}} \stackrel{\text{def}}{=} x^{\text{opt}} - \tilde{x}$ is also small. We can therefore expand the expression for the loss L_0 in Taylor series and keep only terms which are linear and quadratic with respect to Δx. Thus, we get

$$L_0 = u(x^{\text{opt}}, \tilde{a} + \Delta a) - u(x^{\text{opt}} - \Delta x^{\text{opt}}, \tilde{a} + \Delta a)$$

$$= \sum_{i=1}^{n} \frac{\partial u}{\partial x_i}(x^{\text{opt}}, \tilde{a} + \Delta a) \cdot \Delta x_i^{\text{opt}}$$

$$+ \frac{1}{2} \cdot \sum_{i=1}^{n} \sum_{i'=1}^{n} \frac{\partial^2 u}{\partial x_i \partial x_{i'}}(x^{\text{opt}}, \tilde{a} + \Delta a) \cdot \Delta x_i^{\text{opt}} \cdot \Delta x_{i'}^{\text{opt}} + o((\Delta a)^2).$$

By definition, the action x^{opt} maximizes the utility $u(x, \tilde{a} + \Delta a)$. Thus, we have $\frac{\partial u}{\partial x_i}(x^{\text{opt}}, \tilde{a} + \Delta a) = 0$, and the above expression for the loss L_0 takes the simplified form

$$L_0 = \frac{1}{2} \cdot \sum_{i=1}^{n} \sum_{i'=1}^{n} \frac{\partial^2 u}{\partial x_i \partial x_{i'}}(x^{\text{opt}}, \tilde{a} + \Delta a) \cdot \Delta x_i^{\text{opt}} \cdot \Delta x_{i'}^{\text{opt}} + o((\Delta a)^2). \quad (1)$$

The values Δx_i^{opt} can be estimated from the above condition

$$\frac{\partial u}{\partial x_i}(x^{\text{opt}}, \widetilde{a} + \Delta a) = \frac{\partial u}{\partial x_i}(\widetilde{x} + \Delta x^{\text{opt}}, \widetilde{a} + \Delta) = 0.$$

Expanding this expression in Taylor series in terms of Δx_i and Δa_j and taking into account that $\dfrac{\partial u}{\partial x_i}(\widetilde{x}, \widetilde{a}) = 0$ (since for $a = \widetilde{a}$, the utility is maximized by the action $x = \widetilde{x}$), we conclude that for every i, we have

$$\sum_{i'=1}^{n} \frac{\partial^2 u}{\partial x_i \partial x_{i'}}(\widetilde{x}, \widetilde{a}) \cdot \Delta x_{i'}^{\text{opt}} + \sum_{j=1}^{m} \frac{\partial^2 u}{\partial x_i \partial a_j}(\widetilde{x}, \widetilde{a}) \cdot \Delta a_j + o(\Delta x, \Delta a) = 0.$$

Thus, the first approximation Δx_i to the values Δx_i^{opt} can be determined as a solution to a system of linear equations:

$$\sum_{i'=1}^{n} \frac{\partial^2 u}{\partial x_i \partial x_{i'}}(\widetilde{x}, \widetilde{a}) \cdot \Delta x_j = -\sum_{j=1}^{m} \frac{\partial^2 u}{\partial x_i \partial a_j}(\widetilde{x}, \widetilde{a}) \cdot \Delta a_j. \tag{2}$$

A solution to a system of linear equations is a linear combination of the right-hand sides. Thus, the values Δx_i are a linear function of Δa_j. Substituting these linear expressions into the formula (1), we conclude that *the loss L_0 is a quadratic function of Δa_j*, i.e., that $L_0 = \sum_{j=1}^{m} \sum_{j'=1}^{m} k_{jj'} \cdot \Delta a_j \cdot \Delta a_{j'} + o((\Delta a)^2)$ for some coefficients $k_{jj'}$.

Estimating L_1. In the previous section, we considered what happens if we use the original plan $\widetilde{x}$—which was optimal in the original situation $\widetilde{a}$—in the changed situation $a = \widetilde{a} + \Delta a$. Since the original plan is optimal only for the original situation, but not for the new one, using this not-optimal plan leads to the loss L_0, a loss which we estimated as being quadratic in terms of Δa.

To decrease this loss, we need to update the action x. As we have already mentioned, exactly computing the optimal action x^{opt} is, in general, an NP-hard—i.e., computationally intractable—problem. However, as we have also mentioned, the first approximation Δx_i to the desired difference Δx^{opt}—and thus, the first approximation to the newly optimal solution x^{opt}—can be obtained by solving a system of linear equations (2).

The system (2) of linear equations is feasible to solve. Thus, it is reasonable to consider using the action $x^{\text{lin}} = \widetilde{x} + \Delta x$ instead of the original action $\widetilde{x}$. Let us estimate how much we lose if we use this new action x^{lin} instead of the optimal action x_i^{opt}.

The fact that the difference Δx is the first approximation to the optimal difference Δx^{opt} means that we can write $\Delta x^{\text{opt}} = \Delta x + \delta x$, where the remaining term $\delta x \stackrel{\text{def}}{=} \Delta x^{\text{opt}} - \Delta x = x^{\text{opt}} - x^{\text{lin}}$ is of second order in terms of Δx and Δa:

$\delta x = O((\Delta x)^2, (\Delta a)^2)$. Since in the first approximation, Δx has the same order as Δa, we thus get $\delta x = O((\Delta a)^2)$.

The loss L_1 of using $x^{\text{lin}} = x^{\text{opt}} - \delta x$ instead of x^{opt} is equal to the difference

$$L_1 = u(x^{\text{opt}}, \tilde{a} + \Delta a) - u(x^{\text{lin}}, \tilde{a} + \Delta a) = u(x^{\text{opt}}, \tilde{a} + \Delta a) - u(x^{\text{opt}} - \delta x, \tilde{a} + \Delta a).$$

If we expand this expression in δx and keep only linear and quadratic terms, we conclude that

$$L_1 = \sum_{i=1}^{n} \frac{\partial u}{\partial x_i}(x^{\text{opt}}, \tilde{a} + \Delta a) \cdot \delta x_i$$

$$+ \frac{1}{2} \cdot \sum_{i=1}^{n} \sum_{i'=1}^{n} \frac{\partial^2 u}{\partial x_i \partial x_{i'}}(x^{\text{opt}}, \tilde{a} + \Delta a) \cdot \delta x_i \cdot \delta x_{i'} + o((\delta x)^2).$$

Since x^{opt} is the action that, for $a = \tilde{a} + \Delta a$, maximizes utility, we get

$$\frac{\partial u}{\partial x_i}(x^{\text{opt}}, \tilde{a} + \Delta a) = 0.$$

Thus, the expression for L_1 gets a simplified form

$$L_1 = \frac{1}{2} \cdot \sum_{i-1}^{n} \sum_{i'=1}^{n} \frac{\partial^2 u}{\partial x_i \partial x_{i'}}(x^{\text{opt}}, \tilde{a} + \Delta a) \cdot \delta x_i \cdot \delta x_{i'} + o((\delta x)^2).$$

We know that the values δx_i are quadratic in Δa; thus, we conclude that for the modified action, *the loss L_1 is a 4-th order function of Δa_j*, i.e., that

$$L_1 = \sum_{j=1}^{m} \sum_{j'=1}^{m} \sum_{j''=1}^{m} \sum_{j'''=1}^{m} k_{jj'j''j'''} \cdot \Delta a_j \cdot \Delta a_{j'} \cdot \Delta a_{j''} \cdot \Delta a_{j'''} + o((\Delta a)^5)$$

for some coefficients $k_{jj'j''j'''}$.

3 Conclusions

We conclude that:

- the loss L_0 related to using the original plan is quadratic in Δa, while
- the loss L_1 related to using a feasibly modified plan is of 4th order in terms of Δa.

For reasonably small Δa, we have $L_1 \sim (\Delta a)^4 \ll L_0 \sim (\Delta a)^2$.

Let $\varepsilon > 0$ be the maximum loss that we tolerate. Since $L_1 \ll L_0$, we have three possible cases: (1) $\varepsilon < L_1$, (2) $L_1 \le \varepsilon \le L_0$, and (3) $L_0 < \varepsilon$. In the first case, even using the modified action does not help. In the third case, the change in the situation is so small that it is Ok to use the original plan $\tilde{x}$.

In the second case, we have exactly the Eisenhower situation:

- if we use the original plan $\tilde{x}$, the resulting loss L_0 much larger than we can tolerate; in this sense, the original plan is worthless;
- on the other hand, if we feasible modify the original plan into x^{lin}, then we get an acceptable action.

So, we indeed get a theoretical justification of Eisenhower's observation.

Acknowledgements This work was supported in part by the National Science Foundation grants HRD-0734825 and HRD-1242122 (Cyber-ShARE Center of Excellence) and DUE-0926721, and by an award "UTEP and Prudential Actuarial Science Academy and Pipeline Initiative" from Prudential Foundation.

References

1. Eisenhower, D.: A speech to the National Defense Executive Reserve Conference in Washington, D.C., November 14, 1957. In: Eisenhower, D. (ed.) Public Papers of the Presidents of the United States, p. 818. National Archives and Records Service, Government Printing Office (1957)
2. Kreinovich, V., Lakeyev, A., Rohn, J., Kahl, P.: Computational Complexity and Feasibility of Data Processing and Interval Computations. Kluwer, Dordrecht (1998)
3. Luce, R.D., Raiffa, R.: Games and Decisions: Introduction and Critical Survey. Dover, New York (1989)
4. Nguyen, H.T., Kosheleva, O., Kreinovich, V.: Decision making beyond Arrow's 'impossibility theorem', with the analysis of effects of collusion and mutual attraction. Int. J. Intell. Syst. **24**(1), 27–47 (2009)
5. Pardalos, P.: Complexity in Numerical Optimization. World Scientific, Singapore (1993)

Why Convex Optimization Is Ubiquitous and Why Pessimism Is Widely Spread

Angel F. Garcia Contreras, Martine Ceberio and Vladik Kreinovich

Abstract In many practical applications, the objective function is convex. The use of convex objective functions makes optimization easier, but ubiquity of such objective function is a mystery: many practical optimization problems are not easy to solve, so it is not clear why the objective function—whose main goal is to describe our needs—would always describe easier-to-achieve goals. In this paper, we explain this ubiquity based on the fundamental ideas about human decision making. This explanation also helps us explain why in decision making under uncertainty, people often make pessimistic decisions, i.e., decisions based more on the worst-case scenarios.

1 Why Convex Optimization Is Ubiquitous

Reasonable decision making means optimization. In many real life situations, we need to make a decision, i.e., we need to select an alternative x out of many possible alternatives.

Decision making theory has shown that the decision making of a rational person is equivalent to maximizing a special function $u(x)$—known as *utility*—that describes this person's preferences; see, e.g., [1, 5, 6, 8]. Thus, maximization problems are very important for practical applications.

In many cases, the utility value is described by its monetary equivalent amount.

Small changes in an alternative should lead to small change in preferences, so the function $u(x)$ is usually continuous.

A. F. Garcia Contreras · M. Ceberio · V. Kreinovich (✉)
University of Texas at El Paso, El Paso, TX 79968, USA
e-mail: vladik@utep.edu

A. F. Garcia Contreras
e-mail: afgarciacontreras@miners.utep.edu

M. Ceberio
e-mail: mceberio@utep.edu

© Springer Nature Switzerland AG 2020
M. Ceberio and V. Kreinovich (eds.), *Decision Making under Constraints*,
Studies in Systems, Decision and Control 276,
https://doi.org/10.1007/978-3-030-40814-5_12

What if an optimization problem has several solutions? From the purely mathematical viewpoint, it is possible that an optimization problem has several solutions, i.e., several different alternatives $x^{(1)}$, $x^{(2)}$, ... all maximize the objective function $u(x)$:

$$u(x^{(1)}) = u(x^{(2)}) = \cdots = \max_x u(x).$$

From the practical viewpoint, however, the fact that, by using some criterion, we get several possible solutions, means that we can use this non-uniqueness to optimize something else. For example, if a company selects a design x for a new plant, and several designs $x^{(1)}$, $x^{(2)}$, ... are equally profitable, then a reasonable idea is to select, among these most-profitable solutions, the one which is, e.g., the most environmentally friendly. This will weed out some of the possible designs. If even after taking into account environmental impact, we still have several possible alternatives, we can use the remaining non-uniqueness to optimize something else—e.g., look for the most aesthetically pleasing design. This process continues until we end up with the single optimal alternative.

In other words, if the objective function $u(x)$ allows several optimal solutions, this means, from the practical viewpoint, that we need to modify our preferences—i.e., in effect, modify the corresponding objective function—until we end up with an objective function that attains its maximum at the unique point.

So, while, from the mathematical viewpoint, we can consider arbitrary objective functions $u(x)$—and they can serve as good approximations to the way we make decisions—the *final* objective function, the function that describes exactly how we actually make decisions, should have the unique maximum.

How can we describe such final objective functions? In general, selecting a decision x involves selecting the values of many different parameters $x_1, \ldots, x_n$ that characterize this decision. For example, when we select a design of a plant, we must take into account the land area that we need to purchase, the amount of steel and concrete that goes into construction, the overall length of roads, pipes, etc. forming the supporting infrastructure, etc.

Our original decision x is based on known costs of all these attributes. However, costs can change. If the cost per unit of the ith attribute changes by the value d_i, then the overall cost of an option x changes from the original value $u(x)$ to the new value

$$u'(x) = u(x) + \sum_{i=1}^{n} d_i \cdot x_i. \tag{1}$$

It is therefore reasonable to select an objective function $u(x)$ in such away that not only the original function $u(x)$ has the unique maximum, but that for all possible combinations of values d_i, the resulting combination (1) also has the unique maximum.

Need to consider constraints. From the purely mathematical viewpoint, we often consider *unconstrained* optimization, where we have no prior restrictions on the values of the parameters $x_1, \ldots, x_n$ that describe the desired solution $x = (x_1, \ldots, x_n)$. In practice, there are always physical and economical restrictions on the possible values of these parameters. As a result, in practice, for each parameter x_i, we always have bounds $\underline{x}_i$ and $\overline{x}_i$, and we only consider values x_i from the corresponding intervals $[\underline{x}_i, \overline{x}_i]$.

Once we take into account the existence of constraints, we can always guarantee that the corresponding optimization problem always has a solution: indeed, on a bounded closed set $B = [\underline{x}_1, \overline{x}_1] \times \cdots \times [\underline{x}_n, \overline{x}_n]$, every continuous function attains its maximum at some point $x \in B$.

Thus, we arrive at the following definition.

Definition 1 A continuous function $u(x) = u(x_1, \ldots, x_n)$ is called a *final objective function* if for every combination of tuples $d = (d_1, \ldots, d_n)$, $\underline{x} = (\underline{x}_1, \ldots, \underline{x}_n)$, and $\overline{x} = (\overline{x}_1, \ldots, \overline{x}_n)$ the following constrained optimization problem has the unique solution:

$$\text{Maximize } u(x) + \sum_{i=1}^{n} d_i \cdot x_i \text{ under constraints } \underline{x}_i \leq x_i \leq \overline{x}_i.$$

Discussion. There is a class of functions which are realistic objective functions in the sense of the above definition – namely, the class of *strictly convex* functions $u(x)$, i.e., functions for which $u\left(\dfrac{x + x'}{2}\right) > \dfrac{u(x) + u(x')}{2}$ for all $x \neq x'$; see, e.g., [9]. Indeed, it is easy to prove that for a strictly convex function, maximum is attained at a unique point: if we have two different points $x \neq x'$ at which $u(x) = u(x') = \max_x u(x)$, then, due to strong convexity, for the midpoint $x'' \stackrel{\text{def}}{=} \dfrac{x + x'}{2}$, we would have $u(x'') > u(x) = u(x')$, i.e., we would have $u(x'') > \max_x u(x)$, which is not possible.

One can also easily check that if a function $u(x)$ is strictly convex, and if we add a linear expression $\sum_{i=1}^{n} d_i \cdot x_i$ to this function, then the resulting sum $u'(x)$ is also strictly convex. Thus, strictly convex functions are indeed final objective functions in the sense of Definition 1.

Interestingly, if we restrict ourselves to smooth (at least three times differentiable) functions, the opposite is also true: only convex objective functions are final in the sense of the above definition.

Proposition 1 *Every smooth final objective function $u(x)$ is convex.*

Comments.

- This result explains why convex objective functions are ubiquitous in practical applications; see, e.g. [9].
- This result is also good for practical applications since, while optimization in general is NP-hard, feasible algorithms are known for solving convex optimization problem; see, e.g., [4, 7].

Proof of Proposition 1 Let us prove this by contradiction. Let us assume that there exists a smooth final objective function $u(x)$ which is not convex. A smooth function is convex if and only if at all points, its matrix of second derivatives is non-positive definite [9]. Since $u(x)$ is not convex, there exists a point p at which this matrix is not non-negative definite. At this point, the Taylor expansion of the function $u(x)$ has the form

$$u(x) = u(p) + \sum_{i=1}^{n} u_{,i} \cdot (x_i - p_i) + \frac{1}{2} \cdot \sum_{i=1}^{n} \sum_{j=1}^{n} u_{,ij} \cdot (x_i - p_i) \cdot (x_j - p_j) + o((x - p)^2),$$

where $u_{,i} \overset{\text{def}}{=} \dfrac{\partial u}{\partial x_i}$ and $u_{,ij} \overset{\text{def}}{=} \dfrac{\partial^2 u}{\partial x_i \partial x_j}$. Thus, the function $u'(x) = u(x) - \sum_{i=1}^{n} u_{,i} \cdot x_i$ has the form $u'(x) = q(x) + o((x - p)^2)$, where

$$q(x) \overset{\text{def}}{=} u'(p) + \frac{1}{2} \cdot \sum_{i=1}^{n} \sum_{j=1}^{n} u_{,ij} \cdot (x_i - p_i) \cdot (x_j - p_j).$$

Let us take $\underline{x}_i = x_i^{(0)} - \varepsilon$ and $\overline{x}_i = x_i^{(0)} + \varepsilon$ for some small $\varepsilon > 0$. Then, for small $\varepsilon > 0$, $u(x)$ is very close to $q(x)$.

Non-negative definite would mean that $\sum_{i=1}^{n} \sum_{j=1}^{n} u_{,ij} \cdot (x_i - p_i) \cdot (x_j - p_j) \leq 0$ for all x_i. The fact that the matrix $u_{,ij}$ is not non-negative definite means that there exists a vector $x_i - p_i$ for which $\sum_{i=1}^{n} \sum_{j=1}^{n} u_{,ij} \cdot (x_i - p_i) \cdot (x_j - p_j) > 0$. So, for a vector proportional to $x_i - p_i$ and which is within the box B, we have $q(x) > q(p)$. Thus, the maximum of the function $q(x)$ on the box B is *not* attained at p. Since the function $q(x)$ does not change if we reverse the sign of all the differences $x_i - p_i$, with each point $x = p + (x - p)$, the same maximum is attained at a different point $p - (x - p)$. So, for the function $q(x)$, the maximum is attained in at least two different points.

Let us now consider the original function $u'(x)$. If its maximum is attained at two different points, we get our contradiction. Let us now assume that its maximum m is attained at a single point y. This maximum is close to a maximum of the function $q(x)$. The fact that this function has only one maximum means that the value of $u'(x)$ at the point $p - (y - p)$ is slightly smaller than the value $m = u'(y)$. We can then take the plane (linear function) $u = m$, and, keeping its value to be m at the point y, we slightly rotate it and lower it until we touch some other point on the graph— close to $p - (y - p)$. This is possible for $q(x)$, thus it is possible for any function which is sufficiently close to $q(x)$—in particular, for a function $u'(x)$ corresponding to a sufficiently small value $\varepsilon > 0$. Thus, we get a sum $u''(x)$ of $u'(x)$ and a linear function that has at least two maxima. Since $u'(x)$ is itself a sum of $u(x)$ and a linear function, this means that $u''(x)$ is also a sum of $u(x)$ and a linear function—so we get a contradiction with our assumption that the function $u(x)$ is a final objective function.

The proposition is proven.

2   Why Pessimism Is Widely Spread

Decision making under uncertainty. In many practical situations, we do not know the exact consequences of different actions. In other words, for each alternative x, instead of a single value $u(x)$, we have several different values $u(x, s)$ depending on the situation s. According to decision theory, in such situation, a reasonable idea is to optimize the so-called Hurwicz criterion

$$U(x) = \alpha \cdot \max_{s} u(x, s) + (1 - \alpha) \cdot \min_{s} u(x, s)$$

for some $\alpha \in [0, 1]$; see, e.g., [2, 3, 5]. Here, $\alpha = 1$ corresponds to the optimistic approach, when we only consider the best-case scenarios, $\alpha = 0$ is pessimistic approach, when we only consider the worst cases, and $\alpha \in (0, 1)$ means that we consider both the best and the worst cases.

When is this convex? From the viewpoint described in the previous section, it is reasonable to consider situations in which $u(x, s)$ is convex for every s *and* the objective function $U(x)$ is also convex.

For $\alpha = 0$, it is easy to show that the minimum of convex function is always convex; see, e.g., [9]. For $\alpha = 0.5$, we get the arithmetic average which is also convex. For $\alpha < 0.5$, we get a convex combination of cases $\alpha = 0$ and $\alpha = 0.5$, so we also get a convex functions.

However, for any $\alpha > 0.5$, this is no longer true. For example, let us take $s \in \{-, +\}$, with

$$u(x, +) = |x - 1| \text{ and } u(x, -) = |x + 1|.$$

Then, for every $\alpha > 0.5$, the function $U(x)$ attains its maximum value α at two different points. Thus, $U(x)$ is not convex.

This explains why pessimism is widely spread. The fact that only in the pessimistic approach we can guaranteed that the resulting objective function is final explains why the pessimistic approach ($\alpha \leq 0.5$) is widely spread.

Acknowledgements This work was supported in part by the National Science Foundation grants HRD-0734825 and HRD-1242122 (Cyber-ShARE Center of Excellence) and DUE-0926721, and by an award "UTEP and Prudential Actuarial Science Academy and Pipeline Initiative" from Prudential Foundation.

References

1. Fishburn, P.C.: Utility Theory for Decision Making. Wiley, New York (1969)
2. Hurwicz, L.: Optimality Criteria for Decision Making Under Ignorance. Cowles Commission Discussion Paper, Statistics, No. 370 (1951)

3. Kreinovich, V.: Decision making under interval uncertainty (and beyond). In: Guo, P., Pedrycz, W. (eds.) Human-Centric Decision-Making Models for Social Sciences, pp. 163–193. Springer (2014)
4. Kreinovich, V., Lakeyev, A., Rohn, J., Kahl, P.: Computational Complexity and Feasibility of Data Processing and Interval Computations. Kluwer, Dordrecht (1998)
5. Luce, R.D., Raiffa, R.: Games and Decisions: Introduction and Critical Survey. Dover, New York (1989)
6. Nguyen, H.T., Kosheleva, O., Kreinovich, V.: Decision making beyond Arrow's 'impossibility theorem', with the analysis of effects of collusion and mutual attraction. Int. J. Intell. Syst. **24**(1), 27–47 (2009)
7. Pardalos, P.: Complexity in Numerical Optimization. World Scientific, Singapore (1993)
8. Raiffa, H.: Decision Analysis. Addison-Wesley, Reading, MA (1970)
9. Rockafeller, R.T.: Convex Analysis. Princeton University Press, Princeton, NJ (1997)

Probabilistic Fuzzy Neural Networks and Interval Arithmetic Techniques for Forecasting Equities

Ashton Gauff and Humberto Munoz Barona

Abstract We develop an architecture for a neural network specifically for forecasting equities. By implementing the already successful probabilistic fuzzy neural network (PFNN) (developed by Han-Xiong Li and Zhi Liu), we are able to handle complex stochastic uncertainties. In light of the Black Scholes Merton model, we use triangular fuzzy numbers in its calcualtion to include the range instead of a spot price for an investment. Stock market forecasting is multi-dimensional, where time is not the only variable of interest and is non-linear. However, the stochastic gradient descent accounts for optimization in one-dimension for a convex function. We optimize the family of functions with the Hamilton-Jacobi partial differential equation to more accurately represent a non-linear and non-convex function (developed by Chaudhari et al.). In this approach, we will observe the neural network in light of the Dupire formula. Dupire continues the Black Scholes Merton model (BSM), but addresses the weakness of the volatility in the model. By observing volatility as a function of time, Dupire removes the predictable volatility smile found in the BSM model. We optimize this volatility to minimize the risk in forecasting for an optimal prediction. In a future study, we plan to analyze the program's efficiency for forecasting stock options.

1 Mathematical Stochastics

1.1 Brownian Motion

The field of financial asset pricing is highly related from the field of stochastic calculus. The price of a stock tends to follow a Brownian motion (Fig. 1).

A. Gauff · H. Munoz Barona (✉)
Southern University and A&M College, Baton Rouge, LA 70813, USA
e-mail: hmunoz40@hotmail.com

A. Gauff
e-mail: ash70791@gmail.com

© Springer Nature Switzerland AG 2020
M. Ceberio and V. Kreinovich (eds.), *Decision Making under Constraints*,
Studies in Systems, Decision and Control 276,
https://doi.org/10.1007/978-3-030-40814-5_13

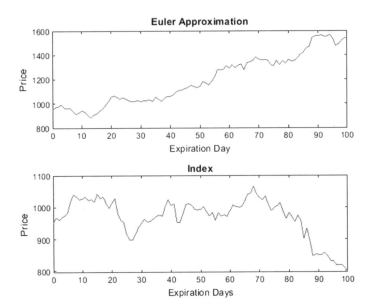

Fig. 1 Forecast euler approximation versus index

Definition 1 A stochastic process $w(t)$ is said to follow a Brownian motion on $[0, T]$ if it satisfies the following:

- $w(0) = 0$
- $w(t)$ is almost continuous.
- For arbitrary $t_1, t_2, \ldots, t_n$, where $0 < t_1 < t_2 < \cdots < t_n < T$, the variables $w(0)$, $w(t_1) - w(0), w(t_2) - w(t_1), \ldots, w(t_n) - w(t_{n-1})$ are mutually independent. In other words, it is a process with **independent increments**.
- The mean (or expected value) is zero, $Ew(t) = 0$.
- The process $w(t)$ takes on a normal distribution density around its mean, more precisely
$$E[w(t_{k+1}) - w(t_k)]^2 = t_{k+1} - t_k.$$

1.2   Ito;s Stochastic Differential and Ito's Lemma

The following definition and theorem are taken from [6].

Definition 2 Suppose $w(t)$ follows a Brownian motion on $[0,T]$, and there exist two functions $u(t)$ and $v(t)$ that are measurable on $[0,T]$, the Ito's stochastic differential of a process $X(t)$ is defined by $dX(t) = u(t)dt + v(t)dw(t).w(t)$ is called a Wiener process.

Definition 3 A process $X(t)$ takes on geometric Brownian motion if its logarithm follows a Brownian motion. Its differential takes on the form

$$dX(t) = aX(t)dt + bX(t)dw(t),\qquad(1)$$

where a and b are constants and $w(t)$ is a Brownian motion.

Theorem 1 (Ito's Lemma) *Suppose that the process $X(t)$ has the stochastic differential $dX(t) = u(t)dt + v(t)dw(t)$ and that the function $f(t, x)$ is nonrandom and is defined for all t and x. Additionally, suppose f is continuous and has continuous derivatives $f_t(t, x)$, $f_x(t, x)$, $f_{xx}(t, x)$. Then the stochastic process $Y(t) = f(t, X(t))$ has also a stochastic differential, and*

$$dY(t) = \left[f_t(t, X(t)) + f_x(t, X(t))u(t) + \frac{1}{2} f_{xx}(t, X(t))v^2(t) \right] dt + f_x(t, X(t))v(t)dw(t)$$
$$(2)$$

or in integral form,

$$Y(t_f) - Y(t_i) = \int_{t_i}^{t_f} \left[f_t(t, X(t)) + f_x(t, X(t))u(t) + \frac{1}{2} f_{xx}(t, X(t))v^2(t) \right] dt$$
$$+ \int_{t_i}^{t_f} f_x(t, X(t))v(t)dw(t)\qquad(3)$$

The Ito's multiplication table for the product of differentials is

$\times$	dw	dt
dw	dt	0
dt	0	0

2 Black-Scholes-Merton Model

The BSM model was developed in the early 1970s for the pricing of securities [5]. It follows a stochastic process as determined by the underlying price with constant volatility and normally distributed returns, where the security pays no dividends. It assumes no transaction costs which is inconsistent with real world activities. There are fees associated with the buying and selling of securities and there are spreads within the bids and asks, allowing for arbitrage to gain from pricing imbalances. The equation does allow for a liquid market where the selling of partial stocks are permitted. In consideration of options, an American option can be bought and sold until its expiration; whereas, an European option must be held until its expiration date. The BSM model is commonly used to determine the price of a European call option and it is described by the following formula

$$C(S(t), t) = S(t)N(d_1) - Ke^{-rt}N(d_2) \tag{4}$$

$$d_1 = \frac{ln\left(\frac{S(t)}{K}\right) + \left(r + \frac{\sigma^2}{2}\right)t}{\sigma\sqrt{t}} \tag{5}$$

$$d_2 = d_1 - \sigma\sqrt{t}, \tag{6}$$

where $C(S(t), t)$ is the price of a call option, K is the strike price of the option, r is the risk free rate of interest, $S(t)$ is the price of the underlying asset, σ is the volatility or standard deviation of the stock's returns, t is the time to expiration, and $N(\cdot)$ is the cummulative standard normal distribution.

3   The Underlying Price

The underlying price $S(t)$ is the current market value of a security with constant returns μ (the lognormal mean is the instantaneous expected rate of change in $S(t)$) and constant volatility σ (the instantaneous standard deviation is the change in security over time). We consider the underlying price function as a function of t and a winner process $w(t)$, $S(t) = S(t, w(t))$ and from (1) its differential is

$$dS(t) = \mu S(t)dt + \sigma S(t)dw(t). \tag{7}$$

Taking $S(t, w(t))$ in it's Taylor expansion following Ito's Lemma, and the Ito's multiplication table it is obtained:

$$\begin{aligned} dS(t) &= \frac{\partial S(t)}{\partial t}dt + \frac{\partial S(t)}{\partial w(t)}dw(t) + \frac{1}{2}\frac{\partial^2 S(t)}{\partial w^2(t)}(dw(t))^2 \\ &+ \frac{\partial^2 S(t)}{\partial w(t)\partial t}(dw(t)dt) + \frac{1}{2}\frac{\partial^2 S(t)}{\partial t^2}(dt)^2 \\ &= \frac{\partial S(t)}{\partial t}dt + \frac{\partial S(t)}{\partial w(t)}dw(t) + \frac{1}{2}\frac{\partial^2 S(t)}{\partial w^2(t)}dt \end{aligned} \tag{8}$$

We treat $w(t)$ as an independent variable, this is to say that the partial derivatives with respect to t treats $w(t)$ as a constant. Following from (7) we have

$$\frac{\partial S(t)}{\partial t} = \mu S(t) \tag{9}$$

$$\frac{\partial S(t)}{\partial w(t)} = \sigma S(t) \tag{10}$$

We can say

$$\frac{\partial^2 S(t)}{\partial w(t)^2} = \frac{\partial}{\partial w(t)} \frac{\partial S(t)}{\partial w(t)} = \frac{\partial}{\partial w(t)} \sigma S(t) = \sigma \frac{\partial}{\partial w(t)} S(t) = \sigma^2 S(t) \quad (11)$$

So that substituting (9), (10), and (11) into (8) we have

$$dS(t) = \mu S(t)dt + \sigma S(t)dw(t) + \frac{1}{2}\sigma^2 S(t)dt \quad (12)$$

Dividing by $S(t)$ and integrating in both sides we have the following:

$$\int \frac{dS}{S(t)} = \int \mu dt + \int \sigma dw(t) + \int \frac{1}{2}\sigma^2 dt$$

$$\ln S(t) = \mu t + \sigma w(t) + \frac{\sigma^2}{2}t + c$$

$$\ln S(t) = \sigma w(t) + \left(\mu + \frac{\sigma^2}{2}\right)t + c$$

$$S(t) = S_0 e^{\sigma w(t) + \left(\mu + \frac{\sigma^2}{2}\right)t} \quad (13)$$

Knowing $S_0 = 1$, we have

$$S(t) = e^{\sigma w(t) + \left(\mu + \frac{\sigma^2}{2}\right)t} \quad (14)$$

4 Dupire Formula

Past models considered the volatility a constant rather than a variable in time [2]. This causes bias in the prediction and pricing of options, where the relationship between the strike and volatility yields the volatility smile. This means that the outcomes of the Black-Scholes equation are predictible and not an accurate representation of the market. Dupire proposed pricing aids in the removal of the volatility smile for a time dependent volatility [3] (Fig. 2).

We have the underlying option given by the following equation [7]:

$$\frac{dS(t)}{S(t)} = \mu(t)dt + \sigma(S(t), t)dw(t), \quad (15)$$

where $\mu(t)$ is the expected return defined by

$$\mu(t) = r(t) - q(t), \quad (16)$$

where $q(t)$ is the dividend, and $r(t)$ is the risk free rate at time t. The differential equation below defines the cummulative distribution function $f(S(t), t)$ for the underlying price $S(t)$ at time t.

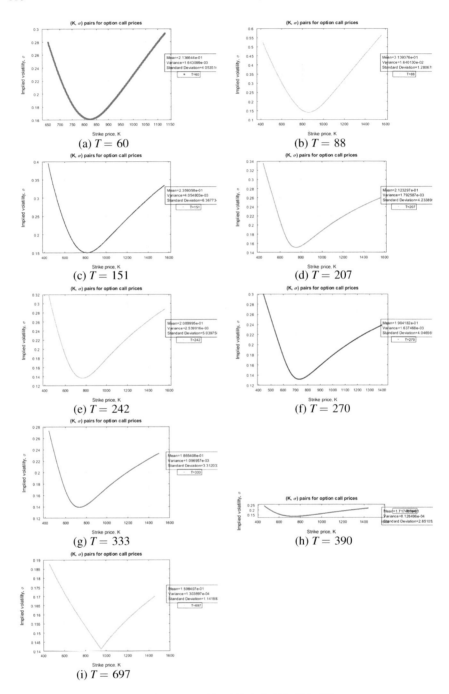

Fig. 2 Volatility smile for T = 60, 88, 151, 207, 242, 270, 333, 390, 697

$$\frac{\partial f}{\partial t} = -\frac{\partial}{\partial S(t)} \left[\mu(t) S(t) f(S(t), t) \right] + \frac{1}{2} \frac{\partial^2}{\partial S(t)^2} \left[\sigma^2(S(t), t) S(t)^2 f(S(t), t) \right].$$

(17)

Lastly, assuming $S(t) > K$, and T is the time at maturity, the European call function is defined by the expected value of $S(t) - K$,

$$C(S(t), T) = \mathbf{E}[S(t) - K] = \int_K^\infty (S(t) - K) f(S(t), T) dS(t)$$

(18)

We begin by finding the derivatives of the call $C(S(t), T)$ with respect to the strike price K.

$$\frac{\partial C(S(t), T)}{\partial K} = \lim_{h \to 0} \frac{\int_{K+h}^\infty (S(t) - (K+h)) f(S(t), T) dS(t) - \int_K^\infty (S(t) - K) f(S(t), T) dS(t)}{h}$$

$$= \lim_{h \to 0} \frac{-h \int_{K+h}^\infty f(S(t), T) dS(t) + \int_{K+h}^\infty (S(t) - K) f(S(t), T) dS(t) - \int_K^\infty (S(t) - K) f(S(t), T) dS(t)}{h}$$

$$= -\int_K^\infty f(S(t), T) dS(t) - \lim_{h \to 0} \frac{\int_K^{K+h} (S(t) - K) f(S(t), T) dS(t)}{h}$$

$$= -\int_K^\infty f(S(t), T) dS(t)$$

(19)

Then the second derivative with respect to K is

$$\frac{\partial^2 C(S(t), T)}{\partial K^2} = f(K, T)$$

(20)

The first derivative of the call $C(S(t), T)$ with respect to $t = T$ assuming no risk at this time.

$$\frac{\partial C(S(t), T)}{\partial T} = \frac{\partial}{\partial T} \left[\int_K^\infty (S(t) - K) f(S(t), T) dS(t) \right]$$

$$= \int_K^\infty (S(t) - K) \frac{\partial}{\partial T} f(S(t), T) dS(t)$$

(21)

From substituting equation (17) and considering $\sigma(S(t), t) = \sigma$ we have

$$\frac{\partial C(S(t), T)}{\partial T} = -\int_K^\infty (S(t) - K) \left[\frac{\partial}{\partial S(t)} [\mu(t) S(t) f(S(t), T)] \right] dS(t)$$

$$+ \frac{1}{2} \int_K^\infty (S(t) - K) \frac{\partial^2}{\partial S(t)^2} \left[\sigma^2 S(t)^2 f(S(t), T) \right] dS(t) \quad (22)$$

Integrating both integrals by parts, we get

$$\frac{\partial C(S(t), T)}{\partial T} = -\left[\mu(t)(S(t) - K)S(t)f(S(t), T)\right]_{S(t)=K}^{S(t)=\infty}$$

$$+\frac{1}{2}\left[(S(t) - K)\frac{\partial}{\partial S(t)}\left(\sigma^2 S(t)^2 f(S(t), T)\right)\right]_{S(t)=K}^{S(t)=\infty}$$

$$+\int_K^\infty \mu(t)S(t)f(S(t), T)dS(t) - \frac{1}{2}\int_K^\infty \frac{\partial}{\partial S(t)}\left[\sigma^2 S(t)^2 f(S(t), T)\right]dS(t)$$

$$= \int_K^\infty \mu(t)S(t)f(S(t), T)dS(t) - \frac{1}{2}\int_K^\infty \frac{\partial}{\partial S(t)}\left[\sigma^2 S(t)^2 f(S(t), T)\right]dS(t)$$

$$= \mu(t)\left[\int_K^\infty (S(t) - K)f(S(t), T)dS(t)\right] + \mu(t)K\int_K^\infty f(S(t), T)dS(t)$$

$$-\frac{1}{2}\int_K^\infty \frac{\partial}{\partial S(t)}\left[\sigma^2 S(t)^2 f(S(t), T)\right]dS(t)$$

$$= \mu(t)\left[C - K\frac{\partial C}{\partial K}\right] + \frac{1}{2}\sigma^2 K^2 \frac{\partial^2 C}{\partial K^2} \tag{23}$$

Assuming no risk, $\mu(t) = 0$, and from (20) we get the differential equation

$$\frac{\partial C(S(t), T)}{\partial T} = \frac{1}{2}\sigma^2 K^2 \frac{\partial^2 C(S(t), T)}{\partial K^2} \tag{24}$$

Solving for σ, we have the Dupire's equation as follows:

$$\sigma(S(t)) = \sqrt{\frac{2\frac{\partial C(S(t),T)}{\partial T}}{K^2 f(K, T)}} \tag{25}$$

5 Dynamic Optimization

To provide a powerful model for the forecasting of securities, we look towards dynamic programming as introduced by Richard Bellman [1]. The provision of a dynamic optimization model divides the problem into smaller problems, where its solutions are combined to yield the robust solution. The principle of optimality holds for all dynamic programming problems, as stated by Bellman: "An optimal policy has the property that whatever the initial state and initial decision are, the remaining decisions must constitute an optimal policy with regard to the state resulting from the first decision."

5.1 Principle of Optimality and Stochastic Optimal Control

Let's consider Itô's stochastic differential equation for a stochastic optimal control problem in discrete time [1],

$$dX(t) = \mu\left(X(t), U(t), t\right) dt + \sigma\left(X(t), U(t), t\right) dw(t),$$

where $X(0) = x_0$. Then we have

$$E\left[\int_0^T F(S(t), U(t), t)dt + S(X(T), T)\right] \quad (26)$$

for a control variable $U(t)$.

Definition 4 (*Principle of Optimality*) For a known value function $V(X(t), t)$, the principle of optimality is defined by

$$V(X(t), t) = \max_u E\left[F(x, u, t)dt + V(X(t) + dX(t), t + dt)\right]. \quad (27)$$

From Ito's Lemma we have

$$V(X(t) + dX(t), t + dt) = V(X(t), t) + V_t dt + V_{X(t)} dX(t) + \frac{1}{2} V_{tX(t)} dt dX(t) + \frac{1}{2} V_{tt} (dt)^2$$
$$+ \frac{1}{2} V_{X(t)X(t)} (dX(t))^2 \quad (28)$$

By the Ito's multiplication rules, we have $(dX(t))^2 = \sigma^2 dt$, $dt dX(t) = 0$, and $(dt)^2 = 0$. Thus we have

$$V(X(t), t) = V(X(t), t)$$
$$+ \max_u E\left[F(x, u, t)dt + V_t(dt) + V_{X(t)}\mu dt + V_{X(t)} + \frac{1}{2}\sigma^2 V_{X(t)X(t)} dt\right] \quad (29)$$

so that we have the following:

$$0 = \max_u E\left[F(x, u, t)dt + V_t(dt) + V_{X(t)}\mu dt + V_{X(t)} + \frac{1}{2}\sigma^2 V_{X(t)X(t)} dt\right]. \quad (30)$$

When we divide by dt we have the Hamilton Jacobi Bellman Equation:

$$0 = \max_u E\left[F(x, u, t) + V_t + V_{X(t)}\mu + V_{X(t)} + \frac{1}{2}\sigma^2 V_{X(t)X(t)}\right]. \quad (31)$$

6 Fuzzification

Consider an stock index where the price $S(t)$, volatility $\sigma(t)$, and drift $\mu(t)$ are known for all beginning time t to $T - t$. We begin by denoting the volatility as a fuzzy triangular number. We must decide how much the price may deviate from the given values. This deviation we will denote as $\lambda_1, \lambda_2 \in [0, 1]$. So that we have each triangular fuzzy number as the following:

$$(\sigma(t)(1 - \lambda_1), \sigma(t), \sigma(t)(1 + \lambda_2)) \tag{32}$$

7 Interval Fuzzy Modeling Algorithm for Optimal Forecast

The Interval fuzzy modeling algorithm for optimal forecast is summarized in this section. The steps in this algorithm implement fuzzy techniques presented in [1, 8, 10–12].

---

1 /* Algorithm as a recursive function of the adaptation for
   an optimal forecast   */
Data: $S(t), K(t), r(t), t$, and $C(t)$.
Result: Minimum volatility or risk σ.
2 **if** *all inputs are given* **then** calculation of the volatility, σ, from Dupire's equation

3 **end if**
4 **if** *volatility $\sigma \leq .01$* **then**
5 | Fuzzification of the volatility into triangular fuzzy numbers giving it a left and right spread to remove arbitrage.
6 **else**

7 **end if**
8 **while** $i < n$ **do**
9 | Calculation of the forecast of black scholes with the given inputs and fuzzy numbers supplied from the fuzzification.
10 | Defuzzification of the results for a singular outcome.
11 **end while**
12 Dynamically optimize the solution by minimizing the volatility or risk to optimize the forecast.

Algorithm 1: Recursive adaptation for an optimal forecast

---

8 Conclusion

This paper suggests an interval fuzzy modeling algorithm for interval time series optimal forecasting. In future study the authors will study the performance and efficiency of this modeling algorithm will be analyzed in forecasting different stock options.

Acknowledgements The authors thank the organizers of the IFSA-NAFIPS 2019 Conference for accepting our research to be presented and to submitted for the post conference proceedings. The authors also thank Dr. Rachael Vincent-Finley Assistant Dean of the Colleague of Sciences and Engineering at SUBR for her support.

References

1. Bellman, R.E.: Mathematics Research Collection: Adaptive Control Processes: A Guided Tour. Princeton Legacy Library. Princeton University Press, New York (2015)
2. Dupire, B.: Pricing with a smile. J. Risk **44**(3), 18–20 (1994)
3. Gatheral, J., Taleb, N.N.: The Volatility Surface: A Practitioner's Guide, 1st edn, Wiley Finance. Wiley, New Jersey (2006)
4. Grzegorzewski, P.: On the Interval Approximation of Fuzzy Numbers. IPMU: Part III, CCSS 299, 59–68, p. 2012. Springer, Heidelberg (2012)
5. Hull, J.: Options, Futures and Other Derivatives, 7th edn, Pearson/Prentice Hall, New York (2009)
6. Ito, K.: On stochastic differential equations. Memoirs. AMS **4**, 1–51 (1951)
7. Kani, I., Dermaan, E., Kama, M.: Trading and Hedging Local Volatility. Goldman Sachs Quantitative Strategies Research Notes. Goldman Sachs, New York (1996)
8. Kreinovich, V., Nguyen, H., Wu, B., Xiang, G.: Computing Statistics Under Interval and Fuzzy Uncertainty: Applications to Computer Science and Engineering. Springer, Heidelberg (2012)
9. Li, H., Liu, Z.: A probabilistic neural-fuzzy learning system for stochastic mode-ling. IEEE Trans. Fuzzy Syst. **16**, 898–908 (2008)
10. Luna, I., Ballini, R.: Adaptive fuzzy system to forecast financial time series volatility. J. Intell. Fuzzy Syst. **23**(1), 27–38 (2012)
11. Luna, I., Ballini, R.: Online estimation of stochastic volatility for asset returns. J. IEEE **23**(1), 27–38 (2012)
12. Maciel, L., Ballini, R., Gomide, F.: Evolving granular analytics for interval time series forecasting. J. Granul. Comput. **1**, 213–224 (2015). https://doi.org/10.1007/s4106601600163

P-Completeness of Testing Solutions of Parametric Interval Linear Systems

Milan Hladík

Abstract We deal with a system of parametric interval linear equations and also with its particular sub-classes defined by symmetry of the constraint matrix. We show that the problem of checking whether a given vector is a solution is a P-complete problem, meaning that there unlikely exists a polynomial closed form arithmetic formula describing the solution set. This is true not only for the general parametric system, but also for the symmetric case with general linear dependencies in the right-hand side. However, we leave as an open problem whether P-completeness concerns also the simplest version of the symmetric solution set with no dependencies in the right-hand side interval vector.

1 Introduction

Let us introduce some notation first. An interval matrix is defined as

$$\mathbf{A} := \{A \in \mathbb{R}^{m \times n}; \; \underline{A} \le A \le \overline{A}\},$$

where $\underline{A}$ and $\overline{A}$, $\underline{A} \le \overline{A}$, are given matrices and the inequality is understood entrywise. The midpoint and radius matrices are defined as

$$A_c := \frac{1}{2}(\underline{A} + \overline{A}), \quad A_\Delta := \frac{1}{2}(\overline{A} - \underline{A}).$$

The set of all interval $m \times n$ matrices is denoted by $\mathbb{IR}^{m \times n}$. Interval vectors are defined analogously.

In this paper, we deal with solutions of parametric interval linear systems. Consider a parametric interval linear system

M. Hladík (✉)
Faculty of Mathematics and Physics, Department of Applied Mathematics,
Charles University, Malostranské nám. 25, 11800 Prague, Czech Republic
e-mail: hladik@kam.mff.cuni.cz

© Springer Nature Switzerland AG 2020
M. Ceberio and V. Kreinovich (eds.), *Decision Making under Constraints*,
Studies in Systems, Decision and Control 276,
https://doi.org/10.1007/978-3-030-40814-5_14

$$A(p)x = b(p),$$

in which parameters have a linear structure

$$A(p) = \sum_{k=1}^{K} A^{(k)} p_k, \quad b(p) = \sum_{k=1}^{K} b^{(k)} p_k.$$

Herein, $A^{(1)}, \ldots, A^{(K)} \in \mathbb{R}^{n \times n}$ and $b^{(1)}, \ldots, b^{(K)} \in \mathbb{R}^n$ are fixed, and the parameters $p_1, \ldots, p_K$ come from their respective interval domains $\mathbf{p}_1, \ldots, \mathbf{p}_K \in \mathbb{IR}$. The solution set of this parametric system is denoted by Σ and consists of all the solutions of all the linear systems corresponding to all possible combinations of parameters $p \in \mathbf{p}$, that is,

$$\Sigma = \{x \in \mathbb{R}^n; \ \exists p \in \mathbf{p} : A(p)x = b(p)\}.$$

As an important subclass, we will also consider a symmetric interval system, where the dependencies determine the constraint matrix to be symmetric. Given $\mathbf{A} \in \mathbb{IR}^{n \times n}$ and $\mathbf{p} \in \mathbb{IR}^K$, the symmetric solution set reads

$$\Sigma_{sym} = \{x \in \mathbb{R}^n; \ \exists p \in \mathbf{p} \ \exists A \in \mathbf{A} : Ax = b(p), \ A = A^T\}. \tag{1}$$

As long as there are no dependencies in the right-hand side, the corresponding symmetric solution set is

$$\Sigma^*_{sym} = \{x \in \mathbb{R}^n; \ \exists A \in \mathbf{A} \ \exists b \in \mathbf{b} : Ax = b, \ A = A^T\}, \tag{2}$$

where $\mathbf{A} \in \mathbb{IR}^{n \times n}$ and $\mathbf{b} \in \mathbb{IR}^n$.

Problem formulation. Given $x^* \in \mathbb{R}^n$, decide whether $x^* \in \Sigma$, or $x^* \in \Sigma_{sym}$, or $x^* \in \Sigma^*_{sym}$.

This problem can be solved by means of linear programming simply considering interval parameters as variables. The decision problems then reduce to checking feasibility of the linear systems

$$A(p)x^* = b(p), \ p \in \mathbf{p}$$

with respect to variables p, and

$$Ax^* = b(p), \ A = A^T, \ A \in \mathbf{A}, \ p \in \mathbf{p}$$

with respect to variables A and p. This can be checked in polynomial time by using linear programming. However, linear programming is a P-complete problem, so that is why we focus on the question whether our problems are P-complete, too.

In contrast, the solution set of the standard interval linear system $Ax = b$, $A \in \mathbf{A}$, $b \in \mathbf{b}$ is defined as

$$\{x \in \mathbb{R}^n;\ \exists A \in \mathbf{A}\ \exists b \in \mathbf{b} : Ax = b\},$$

and characterized by the Oettli–Prager inequalities [12]

$$|A_c x - b_c| \le A_\Delta |x| + b_\Delta.$$

Obviously, checking whether a given point $x^* \in \mathbb{R}^n$ satisfies this system is in NC, which is the class of decision problems decidable in polylogarithmic time on a parallel computer with a polynomial number of processors. Thus, unless P = NC, it is not a P-hard problem.

Since 1990s, there were various attempts to find also a closed form arithmetic description of Σ^*_{sym} and the related solution sets [1–3]. The first approaches utilized the Fourier–Motzkin elimination (yielding possibly double exponential number of constraints). In [6], there was presented an explicit description consisting of an exponential number of constraints, and later improvements include [9–11]. For a general linear parametric solution set Σ, explicit descriptions were presented in [7, 13, 14].

So far, no explicit polynomial Oettli–Prager-type characterization of the solutions of the parametric systems (or some particular subclasses such as the symmetric case) was found. The main message of this paper is to show that such a characterization unlikely exists because the problem is P-complete.

2 Results

Now, we state and prove our main results on P-completeness of testing solutions of parametric systems. We will utilize the fact that checking solvability of a linear system $Ax = b$, $x \ge 0$ is P-complete under NC-reduction (i.e., in polylogarithmic time on a parallel computer with a polynomial number of processors).

First, we state the result for a general parametric system, and the we strengthen it to the symmetric one.

Theorem 1 *Checking $x^* \in \Sigma$ is a P-complete problem.*

Proof Recall that checking solvability of a linear system $Ax = b$, $x \ge 0$ is P-complete, where $A \in \mathbb{R}^{m \times n}$ and $b \in \mathbb{R}^m$. We reduce this system to our problem. First, rewrite the system to

$$\sum_k (A_{*k} e_k^T) x p_k = b, \quad p \in \mathbf{p}, \tag{3}$$

where $e_k = (0, \ldots, 0, 1, 0, \ldots, 0)^T$ is the kth standard unit vector, $\mathbf{p} = [0, c]^n$, and $c > 0$ is sufficiently large. Denoting $e = (1, \ldots, 1)^T$ to be the vector of ones, we have

$(A_{*k}e_k^T)e = A_{*k}$. Thus, the linear system $Ax = b$, $x \geq 0$ is solvable iff $x := e$ is a solution of sysPfThmPcomplGen. Since c can be chosen such that it has a polynomial size with respect to the input size [15], and the transformation is obviously in NC, the reduction is done.

Theorem 2 *Checking $x^* \in \Sigma_{sym}$ is a P-complete problem, even on a subclass of problems where the constraint matrix is tridiagonal with zero diagonal, and the right-hand side vector has at most one parameter in each entry.*

Proof We will again proceed by a reduction from checking solvability of a linear system $Ax = b$, $x \geq 0$, where $A \in \mathbb{R}^{m \times n}$ and $b \in \mathbb{R}^m$. First rewrite the system to an equivalent one

$$A_{*1}x_1 = y^1, \tag{4a}$$

$$y^1 + A_{*2}x_2 = y^2, \tag{4b}$$

$$y^2 + A_{*3}x_3 = y^3, \tag{4c}$$

$$\vdots$$

$$y^{n-1} + A_{*,n}x_n = y^n, \tag{4d}$$

$$y^n = b, \tag{4e}$$

$$x \in [0, \gamma]^n, \tag{4f}$$

where $\gamma > 0$ is again a sufficiently large constant of polynomial size, and $x \in \mathbb{R}^n$, $y^1, \ldots, y^n \in \mathbb{R}^m$ are variables.

We will construct the master constraint matrix of the symmetric system from (1) as a block diagonal matrix consisting of m blocks, each of size $n + 1$. The kth block and the corresponding part of the right-hand side read

$$\begin{pmatrix} 0 & y_k^1 & 0 & \cdots & & 0 \\ y_k^1 & 0 & -y_k^2 & \ddots & & \vdots \\ 0 & -y_k^2 & 0 & & & 0 \\ \vdots & \ddots & \ddots & \ddots & & (-1)^{n-1}y_k^n \\ 0 & \cdots & 0 & (-1)^{n-1}y_k^n & 0 \end{pmatrix}, \begin{pmatrix} a_{k1}x_1 \\ -a_{k2}x_2 \\ \vdots \\ (-1)^{n-1}a_{kn}x_n \\ (-1)^{n-1}b_k \end{pmatrix}. \tag{5}$$

The equations associated to this block correspond to the kth equations in each of the equalities in (4). As a consequence, (4) is solvable iff the vector of ones belongs to the symmetric solution set obtained from (5); the variables x in (5) correspond to the parameters p in Σ_{sym}, $\mathbf{p} = [0, \gamma]^n$, and the variables y_i^j correspond to the entries of the symmetric matrix from Σ_{sym} and ranges in $[-\gamma, \gamma]$.

We did not succeeded in our attempts to prove P-completeness of testing $x^* \in \Sigma_{sym}^*$, which remains as an open problem:

Open problem: Is the problem of checking $x^* \in \Sigma_{sym}^*$ P-complete?

Nevertheless, we can state some observations for the modified problem, and show relations to other open problems.

Even though complexity of checking $x^* \in \Sigma^*_{sym}$ is unknown, the problem of finding the best certificate is P-complete. Herein, "the best" means that it maximizes the given linear function $\mathrm{tr}(AF)$, where $\mathrm{tr}(\cdot)$ stands for the matrix trace.

Theorem 3 *The following problem is P-complete: Given $A \in \mathbb{R}^{n \times n}$, $b \in \mathbb{IR}^n$, symmetric $F \in \mathbb{R}^{n \times n}$, and $x^* \in \mathbb{R}^n$, among symmetric matrices $A \in \mathbf{A}$ for which $Ax^* \in b$, find the matrix for which $\mathrm{tr}(AF)$ is the largest possible.*

Proof We use the fact that finding the minimal value of a transportation problem

$$\min \sum_{i,j} c_{ij} x_{ij}$$

$$\text{subject to } \sum_{j} x_{ij} \le a_i,$$

$$\sum_{i} x_{ij} \le b_j,$$

$$x_{ij} \ge 0$$

is P-complete [4, 5, 8]. Now, we put $x^* := e$ and construct the master matrix A and the right-hand side b having the form

$$A = \begin{pmatrix} 0 & B \\ B^T & D \end{pmatrix}, \quad b = \begin{pmatrix} d \\ 0 \end{pmatrix}.$$

The first part of the equations $Ax^* = b$ corresponds to the constraints of the transportation problem, and the second part contains auxiliary equations. The right-hand side in particular has the structure

$$d \in \mathbf{d} = \begin{pmatrix} [-\gamma e, a] \\ [b, \gamma e] \end{pmatrix},$$

where γ is a sufficiently large constant as in the above proofs. The nonzero entries of B have the interval domain $[0, \gamma]$ and those of D range in $[-\gamma, \gamma]$. The sub-matrix B is created as follows

$$B = \begin{pmatrix} x_{11} & |x_{12} & |\cdots & \\ x'_{11} & & & \\ & x'_{12} & & \\ & & |x_{mn} & \\ & & |\cdots & x'_{mn} \end{pmatrix}.$$

Each row has the variables (the empty positions are zeros) corresponding to the variables in the transportation problem constraints. Each pair of the depicted columns is associated with one particular variable. In order that two instances x_{ij} and x'_{ij} of one variable have the same value, the matrix D is constructed as block diagonal with blocks of size two

$$
D = \begin{pmatrix}
0 & w_{11} & & & \\
w_{11} & 0 & & & \\
& & \ddots & & \\
& & & 0 & w_{mn} \\
& & & w_{mn} & 0
\end{pmatrix}.
$$

In this way, the corresponding two equations

$$
x_{ij} + w_{ij} = 0, \quad x'_{ij} + w_{ij} = 0
$$

from $B^T e + De = 0$ give rise to $x_{ij} = x'_{ij}$. Eventually, F has a similar structure as A such that the entries corresponding to variables are set up to $-c_{ij}$, and they are zero otherwise. In this way, we reduced the transportation problem to finding maximal $\mathrm{tr}(AF)$ on the set of symmetric matrices $A \in \mathbf{A}$ satisfying $Ax^* = b \in \mathbf{b}$.

In [5, 8], there is mentioned an open problem of checking solvability of an inequality system such that each constraint involves at most two variables. If the dual problem (each variable is involved in at most two constraints) turns out to be P-complete, then modifying the proof of Theorem 3 also the problem $x^* \in \Sigma^*_{sym}$ will be P-complete.

Another related problem is the homologous flow problem [5], in which we are seeking for a feasible solution of an edge-capacitated network flow problem with homologous edges (pairs of edges required to have the same flow). It is also a P-complete problem, and if it was P-complete also without the homology condition, then we could easily reduce P-completeness of the problem $x^* \in \Sigma^*_{sym}$ as well.

3   Conclusion

We proved P-completeness of checking whether a vector is a solution of a parametric interval linear system of equations, and we strengthened the result to the symmetric case with general linear dependencies in the right-hand side only. We did not succeeded in our attempts to prove P-completeness for the simpler version of the symmetric solution set, which remains as an open problem. Nevertheless, we showed some interesting relations to other open problems as well as P-completeness of the optimization version of the problem in question.

Acknowledgements The author was supported by the Czech Science Foundation Grant P403-18-04735S.

References

1. Alefeld, G., Kreinovich, V., Mayer, G.: On the shape of the symmetric, persymmetric, and skew-symmetric solution set. SIAM J. Matrix Anal. Appl. **18**(3), 693–705 (1997)
2. Alefeld, G., Kreinovich, V., Mayer, G.: On the solution sets of particular classes of linear interval systems. J. Comput. Appl. Math. **152**(1–2), 1–15 (2003)
3. Alefeld, G., Mayer, G.: On the symmetric and unsymmetric solution set of interval systems. SIAM J. Matrix Anal. Appl. **16**(4), 1223–1240 (1995)
4. Goldschlager, L.M., Shaw, R.A., Staples, J.: The maximum flow problem is log space complete for P. Theor. Comput. Sci. **21**, 105–111 (1982)
5. Greenlaw, R., Hoover, H.J., Ruzzo, W.L.: Limits to Parallel Computation: P-Completeness Theory. Oxford University Press, New York (1995)
6. Hladík, M.: Description of symmetric and skew-symmetric solution set. SIAM J. Matrix Anal. Appl. **30**(2), 509–521 (2008)
7. Hladík, M.: Enclosures for the solution set of parametric interval linear systems. Int. J. Appl. Math. Comput. Sci. **22**(3), 561–574 (2012)
8. Lueker, G.S., Megiddo, N., Ramachandran, V.: Linear programming with two variables per inequality in poly-log time. In: Proceedings of the Eighteenth Annual ACM Symposium on Theory of Computing, STOC '86, pp. 196–205. ACM, New York (1986)
9. Mayer, G.: An Oettli-Prager-like theorem for the symmetric solution set and for related solution sets. SIAM J. Matrix Anal. Appl. **33**(3), 979–999 (2012)
10. Mayer, G.: A survey on properties and algorithms for the symmetric solution set. Technical Report 12/2, Universität Rostock, Institut für Mathematik (2012). http://ftp.math.uni-rostock.de/pub/preprint/2012/pre12_02.pdf
11. Mayer, G.: Three short descriptions of the symmetric and of the skew-symmetric solution set. Linear Algebra Appl. **475**, 73–79 (2015)
12. Oettli, W., Prager, W.: Compatibility of approximate solution of linear equations with given error bounds for coefficients and right-hand sides. Numer. Math. **6**, 405–409 (1964)
13. Popova, E.D.: Explicit description of AE solution sets for parametric linear systems. SIAM J. Matrix Anal. Appl. **33**(4), 1172–1189 (2012)
14. Popova, E.D.: Solvability of parametric interval linear systems of equations and inequalities. SIAM J. Matrix Anal. Appl. **36**(2), 615–633 (2015)
15. Schrijver, A.: Theory of Linear and Integer Programming. Repr. Wiley, Chichester (1998)

Thick Separators

Luc Jaulin and Benoit Desrochers

Abstract If an interval of $\mathbb{R}$ is an uncertain real number, a *thick set* is an uncertain subset of $\mathbb{R}^n$. More precisely, a thick set is an interval of the powerset of $\mathbb{R}^n$ equipped with the inclusion $\subset$ as an order relation. It can generally be defined by parameters or functions which are not known exactly, but are known to belong to some intervals. In this paper, we show how to use constraint propagation methods in order to compute efficiently an inner and an outer approximations of a thick set. The resulting inner/outer contraction are made using an operator which is called a *thick separator*. Then, we show how thick separators can be combined in order to compute with thick sets.

1 Thick Sets

A thin set is a subset of $\mathbb{R}^n$. It is qualified as *thin* because its boundary is thin. In this section, we present the definition of thick sets.

Thick set. Denote by $(\mathcal{P}(\mathbb{R}^n), \subset)$, the powerset of $\mathbb{R}^n$ equipped with the inclusion $\subset$ as an order relation. A *thick set* $[\![\mathbb{X}]\!]$ of $\mathbb{R}^n$ is an interval of $(\mathcal{P}(\mathbb{R}^n), \subset)$. If $[\![\mathbb{X}]\!]$ is a thick set of $\mathbb{R}^n$, there exist two subsets of $\mathbb{R}^n$, called the *subset bound* and the *supset bound* such that

$$[\![\mathbb{X}]\!] = \qquad [\mathbb{X}^{\subset}, \mathbb{X}^{\supset}] = \{\mathbb{X} \in \mathcal{P}(\mathbb{R}^n) | \mathbb{X}^{\subset} \subset \mathbb{X} \subset \mathbb{X}^{\supset}\}. \qquad (1)$$

Another representation for the thickset $[\![\mathbb{X}]\!]$ is the partition $\{\mathbb{X}^{in}, \mathbb{X}^?, \mathbb{X}^{out}\}$, where

L. Jaulin (✉) · B. Desrochers
Lab-STICC, ENSTA Bretagne, Brest, France
e-mail: luc.jaulin@gmail.com

B. Desrochers
e-mail: benoit.desrochers@ensta-bretagne.org

© Springer Nature Switzerland AG 2020
M. Ceberio and V. Kreinovich (eds.), *Decision Making under Constraints*,
Studies in Systems, Decision and Control 276,
https://doi.org/10.1007/978-3-030-40814-5_15

$$\begin{aligned}
\mathbb{X}^{in} &= \mathbb{X}^{\complement}\\
\mathbb{X}^? &= \mathbb{X}^{\supset}\backslash\mathbb{X}^{\complement} \\
\mathbb{X}^{out} &= \overline{\mathbb{Z}^{\supset}}.
\end{aligned} \tag{2}$$

The subset $\mathbb{X}^?$ is called the *penumbra* and plays an important role in the characterization of thick sets [3]. Thick sets can be used to represent uncertain sets (such as an uncertain map [4]) or a soft constraints [1].

2 Thick Separators

To characterize a thin set using a paver, we may use a separator (which is a pair of two contractors [6]) inside a paver. Separators can be immediately generalized to thick sets. Now, the penumbra as a non-zero volume for thick sets. For efficiency reasons, it is important to avoid any accumulation of the paving deep inside the penumbra. This is the role of thick separators to avoid as much as possible bisections inside the penumbra.

Thick separators. A *thick separator* $[\![\mathcal{S}]\!]$ for the thick set $[\![\mathbb{X}]\!]$ is an extension of the concept of separator to thick sets. More precisely, a thick separator is a 3-uple of contractors $\{\mathcal{S}^{in}, \mathcal{S}^?, \mathcal{S}^{out}\}$ such that, for all $[\mathbf{x}] \in \mathbb{IR}^n$

$$\begin{aligned}
\mathcal{S}^{in}([\mathbf{x}]) \cap \mathbb{X}^{in} &= [\mathbf{x}] \cap \mathbb{X}^{in}\\
\mathcal{S}^?([\mathbf{x}]) \cap \mathbb{X}^? &= [\mathbf{x}] \cap \mathbb{X}^? \\
\mathcal{S}^{out}([\mathbf{x}]) \cap \mathbb{X}^{out} &= [\mathbf{x}] \cap \mathbb{X}^{out}
\end{aligned} \tag{3}$$

In what follow, we define an algebra for thick separator in a similar manner than what as been done for contractors [2] or for separators [7].

3 Algebra

In this section, we show how we can define operations for thick sets (as a union, intersection, difference, etc.). The main motivation is to be able to compute with thick sets.

Intersection. Consider two thick sets $[\![\mathbb{X}]\!] = [\![\mathbb{X}^{\complement}, \mathbb{X}^{\supset}]\!]$ and $[\![\mathbb{Y}]\!] = [\![\mathbb{Y}^{\complement}, \mathbb{Y}^{\supset}]\!]$ with thick separators $[\![\mathcal{S}_{\mathbb{X}}]\!] = \{\mathcal{S}_{\mathbb{X}}^{in}, \mathcal{S}_{\mathbb{X}}^?, \mathcal{S}_{\mathbb{X}}^{out}\}$ and $[\![\mathcal{S}_{\mathbb{Y}}]\!] = \{\mathcal{S}_{\mathbb{Y}}^{in}, \mathcal{S}_{\mathbb{Y}}^?, \mathcal{S}_{\mathbb{Y}}^{out}\}$. A thick separator for the thick set

$$[\![\mathbb{Z}]\!] = [\![\mathbb{Z}^{\complement}, \mathbb{Z}^{\supset}]\!] = [\![\mathbb{X}]\!] \cap [\![\mathbb{Y}]\!] = [\![\mathbb{X}^{\complement} \cap \mathbb{Y}^{\complement}, \mathbb{X}^{\supset} \cap \mathbb{Y}^{\supset}]\!] \tag{4}$$

is

$$
\begin{aligned}
[\![\mathcal{S}_{\mathbb{Z}}]\!] &= \left\{\mathcal{S}_{\mathbb{Z}}^{in}, \mathcal{S}_{\mathbb{Z}}^{?}, \mathcal{S}_{\mathbb{Z}}^{out}\right\} \\
&= \left\{\mathcal{S}_{\mathbb{X}}^{in} \cap \mathcal{S}_{\mathbb{Y}}^{in}, \left(\mathcal{S}_{\mathbb{X}}^{?} \cap \mathcal{S}_{\mathbb{Y}}^{in}\right) \sqcup \left(\mathcal{S}_{\mathbb{X}}^{?} \cap \mathcal{S}_{\mathbb{Y}}^{?}\right) \sqcup \left(\mathcal{S}_{\mathbb{X}}^{in} \cap \mathcal{S}_{\mathbb{Y}}^{?}\right), \mathcal{S}_{\mathbb{X}}^{out} \sqcup \mathcal{S}_{\mathbb{Y}}^{out}\right\}.
\end{aligned} \tag{5}
$$

Proof. We have

$$
\begin{aligned}
\mathbb{Z}^{in} = \mathbb{Z}^{\mathbb{C}} &= \mathbb{X}^{\mathbb{C}} \cap \mathbb{Y}^{\mathbb{C}} \\
&= \mathbb{X}^{in} \cap \mathbb{Y}^{in} \\
\mathbb{Z}^{?} = \mathbb{Z}^{\supset}\backslash\mathbb{Z}^{\mathbb{C}} &= \mathbb{X}^{\supset} \cap \mathbb{Y}^{\supset}\backslash(\mathbb{X}^{\mathbb{C}} \cap \mathbb{Y}^{\mathbb{C}}) \\
&= \mathbb{X}^{\supset} \cap \mathbb{Y}^{\supset} \cap \overline{\mathbb{X}^{\mathbb{C}} \cap \mathbb{Y}^{\mathbb{C}}} \\
&= \mathbb{X}^{\supset} \cap \mathbb{Y}^{\supset} \cap \left(\overline{\mathbb{X}^{\mathbb{C}}} \cup \overline{\mathbb{Y}^{\mathbb{C}}}\right) \\
&= \left(\mathbb{X}^{in} \cup \mathbb{X}^{?}\right) \cap \left(\mathbb{Y}^{in} \cup \mathbb{Y}^{?}\right) \cap \left(\left(\mathbb{X}^{out} \cup \mathbb{X}^{?}\right) \cup \left(\mathbb{Y}^{out} \cup \mathbb{Y}^{?}\right)\right) \\
&= \left(\mathbb{X}^{?} \cap \mathbb{Y}^{in}\right) \cup \left(\mathbb{X}^{?} \cap \mathbb{Y}^{?}\right) \cup \left(\mathbb{X}^{in} \cap \mathbb{Y}^{?}\right) \\
\mathbb{Z}^{out} = \overline{\mathbb{Z}^{\supset}} &= \overline{\mathbb{X}^{\supset} \cap \mathbb{Y}^{\supset}} \\
&= \overline{\mathbb{X}^{\supset}} \cup \overline{\mathbb{Y}^{\supset}} \\
&= \mathbb{X}^{out} \cup \mathbb{Y}^{out}.
\end{aligned}
$$

From the separator algebra, we get that a contractor for $\mathbb{Z}^{in}$ is $\mathcal{S}_{\mathbb{Z}}^{in} = \mathcal{S}_{\mathbb{X}}^{in} \cap \mathcal{S}_{\mathbb{Y}}^{in}$, a contractor for $\mathbb{Z}^{out}$ is $\mathcal{S}_{\mathbb{Z}}^{out} = \mathcal{S}_{\mathbb{X}}^{out} \sqcup \mathcal{S}_{\mathbb{Y}}^{out}$ and a contractor for $\mathbb{Z}^{?}$ is

$$
\mathcal{S}_{\mathbb{Z}}^{?} = \left(\mathcal{S}_{\mathbb{X}}^{?} \cap \mathcal{S}_{\mathbb{Y}}^{in}\right) \sqcup \left(\mathcal{S}_{\mathbb{X}}^{?} \cap \mathcal{S}_{\mathbb{Y}}^{?}\right) \sqcup \left(\mathcal{S}_{\mathbb{X}}^{in} \cap \mathcal{S}_{\mathbb{Y}}^{?}\right). \tag{6}
$$

4   Using Karnaugh Map

This expression could have been obtained using Fig. 1.

Karnaugh map, as illustrated by Fig. 2, can also can be used to get the expression for thick separators a more clear manner. For instance, if

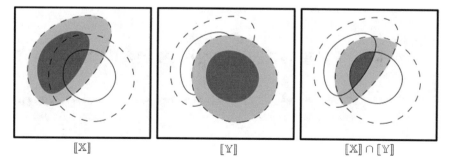

Fig. 1 Intersection of two thick sets. Red means *inside*, Blue means *outside* and Orange means *uncertain*

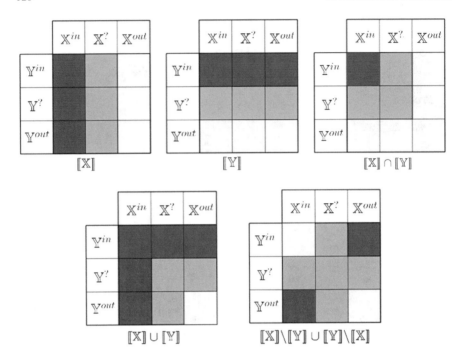

Fig. 2 Karnaugh map

$$[\![\mathbb{Z}]\!] = [\![\mathbb{X}]\!] \cup [\![\mathbb{Y}]\!], \tag{7}$$

we read from the Karnaugh map

$$\begin{array}{rl}
\mathbb{Z}^{in} = & \mathbb{X}^{in} \cup \mathbb{Y}^{in} \\
\mathbb{Z}^? = & \left(\mathbb{X}^? \cap \mathbb{Y}^{out}\right) \cup \left(\mathbb{X}^? \cap \mathbb{Y}^?\right) \cup \left(\mathbb{X}^{out} \cap \mathbb{Y}^?\right) \\
\mathbb{Z}^{out} = & \mathbb{X}^{out} \cap \mathbb{Y}^{out}.
\end{array} \tag{8}$$

Therefore a thick separator for the thick set $[\![\mathbb{Z}]\!]$ is

$$\begin{array}{rl}
[\![\mathcal{S}_{\mathbb{Z}}]\!] = & \left\{\mathcal{S}_{\mathbb{Z}}^{in}, \mathcal{S}_{\mathbb{Z}}^?, \mathcal{S}_{\mathbb{Z}}^{out}\right\} \\
= & \left\{\mathcal{S}_{\mathbb{X}}^{in} \sqcup \mathcal{S}_{\mathbb{Y}}^{in}, \left(\mathcal{S}_{\mathbb{X}}^? \cap \mathcal{S}_{\mathbb{Y}}^{out}\right) \sqcup \left(\mathcal{S}_{\mathbb{X}}^? \cap \mathcal{S}_{\mathbb{Y}}^?\right) \sqcup \left(\mathcal{S}_{\mathbb{X}}^{out} \cap \mathcal{S}_{\mathbb{Y}}^?\right), \mathcal{S}_{\mathbb{X}}^{out} \cap \mathcal{S}_{\mathbb{Y}}^{out}\right\}
\end{array} \tag{9}$$

Now, if

$$[\![\mathbb{Z}]\!] = [\![\mathbb{X}]\!] \backslash [\![\mathbb{Y}]\!] \cup [\![\mathbb{Y}]\!] \backslash [\![\mathbb{X}]\!], \tag{10}$$

we read

$$\begin{array}{rl}
\mathbb{Z}^{in} = & \left(\mathbb{X}^{in} \cap \mathbb{Y}^{out}\right) \cup \left(\mathbb{X}^{out} \cap \mathbb{Y}^{in}\right) \\
\mathbb{Z}^? = & \mathbb{X}^? \cup \mathbb{Y}^? \\
\mathbb{Z}^{out} = & \left(\mathbb{X}^{in} \cap \mathbb{Y}^{in}\right) \cup \left(\mathbb{X}^{out} \cap \mathbb{Y}^{out}\right).
\end{array} \tag{11}$$

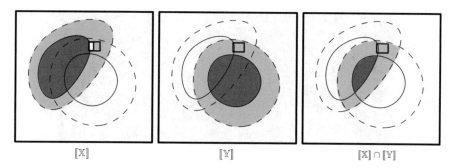

$[\mathrm{X}]$ $[\mathrm{Y}]$ $[\mathrm{X}] \cap [\mathrm{Y}]$

Fig. 3 Illustration of the intersection of two separators

Therefore a thick separator for the thick set $[\![\mathbb{Z}]\!]$ is

$$
\begin{aligned}
[\![\mathcal{S}_{\mathbb{Z}}]\!] &= \left\{ \mathcal{S}_{\mathbb{Z}}^{in}, \mathcal{S}_{\mathbb{Z}}^{?}, \mathcal{S}_{\mathbb{Z}}^{out} \right\} \\
&= \left\{ \mathcal{S}_{\mathbb{X}}^{in} \sqcup \mathcal{S}_{\mathbb{Y}}^{in}, \left(\mathcal{S}_{\mathbb{X}}^{?} \cap \mathcal{S}_{\mathbb{Y}}^{out} \right) \sqcup \left(\mathcal{S}_{\mathbb{X}}^{?} \cap \mathcal{S}_{\mathbb{Y}}^{?} \right) \sqcup \left(\mathcal{S}_{\mathbb{X}}^{out} \cap \mathcal{S}_{\mathbb{Y}}^{?} \right), \mathcal{S}_{\mathbb{X}}^{out} \cap \mathcal{S}_{\mathbb{Y}}^{out} \right\}.
\end{aligned} \quad (12)
$$

Note that when we build such an expression from a Karnaugh map, fake boundaries may appear. They could be avoided using the method proposed in [9].

Example. Take one box $[\mathbf{x}]$ as in Fig. 3. We get

$$
[\![\mathcal{S}_{\mathbb{X}}]\!] ([\mathbf{x}]) = \left\{ \mathcal{S}_{\mathbb{X}}^{in}, \mathcal{S}_{\mathbb{X}}^{?}, \mathcal{S}_{\mathbb{X}}^{out} \right\} ([\mathbf{x}]) = \{[\mathbf{a}], [\mathbf{x}], \emptyset\} \quad (13)
$$

where $[\mathbf{a}]$ the white box. Moreover,

$$
[\![\mathcal{S}_{\mathbb{Y}}]\!] ([\mathbf{x}]) = \left\{ \mathcal{S}_{\mathbb{Y}}^{in}, \mathcal{S}_{\mathbb{Y}}^{?}, \mathcal{S}_{\mathbb{Y}}^{out} \right\} ([\mathbf{x}]) = \{\emptyset, [\mathbf{x}], \emptyset\}. \quad (14)
$$

Thus

$$
\begin{aligned}
[\![\mathcal{S}_{\mathbb{Z}}]\!] &= \left\{ \mathcal{S}_{\mathbb{Z}}^{in}, \mathcal{S}_{\mathbb{Z}}^{?}, \mathcal{S}_{\mathbb{Z}}^{out} \right\} ([\mathbf{x}]) \\
&= \{ \quad \mathcal{S}_{\mathbb{X}}^{in} \cap \mathcal{S}_{\mathbb{Y}}^{in}([\mathbf{x}]), \\
&= \quad \left(\mathcal{S}_{\mathbb{X}}^{?} \cap \mathcal{S}_{\mathbb{Y}}^{in} \right) \sqcup \left(\mathcal{S}_{\mathbb{X}}^{?} \cap \mathcal{S}_{\mathbb{Y}}^{?} \right) \sqcup \left(\mathcal{S}_{\mathbb{X}}^{in} \cap \mathcal{S}_{\mathbb{Y}}^{?} \right) ([\mathbf{x}]), \\
&= \quad \mathcal{S}_{\mathbb{X}}^{out} \sqcup \mathcal{S}_{\mathbb{Y}}^{out}([\mathbf{x}]) \quad \} \\
&= \{ [\mathbf{a}] \cap \emptyset, ([\mathbf{x}] \cap \emptyset) \sqcup ([\mathbf{x}] \cap [\mathbf{x}]) \sqcup ([\mathbf{a}] \cap [\mathbf{x}]), \emptyset \sqcup \emptyset \} \\
&= \quad \{\emptyset, [\mathbf{x}], \emptyset\}
\end{aligned}
$$

We conclude that $[\mathbf{x}] \subset \mathbb{Z}^{in}$.

5 Test Case

Interval linear system [10, 11] are linear systems of equations the coefficients of which are uncertain and belong to some intervals. Consider for instance the following interval linear system [8]:

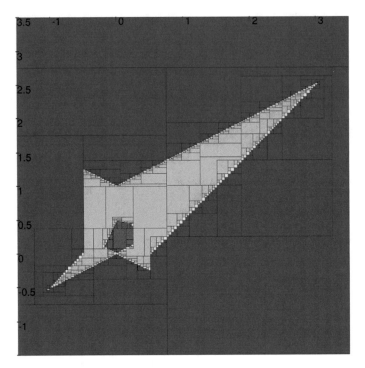

Fig. 4 Thick set corresponding to the test-case. Red boxes are inside the thick solution set and the blue boxes are outside. The penumbra corresponds to the orange boxes

$$\begin{cases} [2, 4] \cdot x_1 + [-2, 0] \cdot x_2 \in [-1, 1] \\ [-1, 1] \cdot x_1 + [2, 4] \cdot x_2 \in [0, 2] \end{cases} \tag{15}$$

For each constraint, a thick separator can be build and then combined using Eq. 5. A thick set inversion algorithm provides the paving Fig. 4. The solution set $[\![\mathbb{X}]\!] = [\![\mathbb{X}^{\subset}, \mathbb{X}^{\supset}]\!]$ has for supset bound the *tolerable solution set* $\mathbb{X}^{\supset}$(red+orange) and for subset bound $\mathbb{X}^{\subset}$ the *united solution set* (red) [5]. Note that inside the penumbra, no accumulation can be observed.

References

1. Brefort, Q., Jaulin, L., Ceberio, M., Kreinovich, V.: If we take into account that constraints are soft, then processing constraints becomes algorithmically solvable. In: Proceedings of the IEEE Series of Symposia on Computational Intelligence SSCI'2014. Orlando, Florida, 9–12 Dec 2014
2. Chabert, G., Jaulin, L.: Contractor programming. Artif. Intell. **173**, 1079–1100 (2009)
3. Desrochers, B., Jaulin, L.: Computing a guaranteed approximation the zone explored by a robot. IEEE Trans. Autom. Control (2016)

4. Desrochers, B., Lacroix, S., Jaulin, L.: Set-membership approach to the kidnapped robot problem. In: IROS 2015 (2015)
5. Goldsztejn, A., Chabert, G.: On the approximation of linear ae-solution sets. In: 12th International Symposium on Scientific Computing, Computer Arithmetic and Validated Numerics,
   Duisburg, Germany, (SCAN 2006) (2006)
6. Jaulin, L., Desrochers, B.: Robust localisation using separators. In: COPROD 2014 (2014)
7. Jaulin, L., Desrochers, B.: Introduction to the algebra of separators with application to path
   planning. Eng. Appl. Artif. Intell. **33**, 141–147 (2014)
8. Kreinovich, V., Shary, S.: Interval methods for data fitting under uncertainty: a probabilistic
   treatment. Reliab. Comput. (2016)
9. Schvarcz Franco, G., Jaulin, L.: How to avoid fake boundaries in contractor programming. In:
   SWIM'16 (2016)
10. Shary, S.: On optimal solution of interval linear equations. SIAM J. Numer. Anal. **32**(2), 610–
    630 (1995)
11. Shary, S.: A new technique in systems analysis under interval uncertainty and ambiguity. Reliab.
    Comput. **8**, 321–418 (2002)

Using Constraint Propagation for Cooperative UAV Localization from Vision and Ranging

Ide-Flore Kenmogne, Vincent Drevelle and Eric Marchand

Abstract This paper addresses the problem of cooperative localization in a group of unmanned aerial vehicles (UAV) in a bounded error context. The UAVs are equipped with cameras to tracks landmarks, and a communication and ranging system to cooperate with their neighbours. Measurements are represented by intervals, and constraints are expressed on the robots poses (positions and orientations). Each robot first computes a pose domain using only its sensors measurements, by using set inversion via interval analysis (Moore in Interval analysis. Prentice Hall, 1966 [1]). Then, through position boxes exchange, positions are cooperatively refined by constraint propagation in the group. Results are presented with real robot data, and show position accuracy improvement thanks to cooperation.

Keywords Intervals · Cooperative localization · Constraints propagation

1 Introduction

In this paper, we consider the problem of cooperative localization [2] in a group of N unmanned aerial vehicles (UAV). The robots are equipped with cameras, able to see landmarks of known positions. A communication and ranging system provides to each robot R_k a means of exchanging data and measuring distances with its neighbours and a base station B (Fig. 1). The goal for each robot is to compute a domain for its pose (position and orientation), assuming bounded error measurements.

The paper is organized as follows: we first present how each robot is able to independently compute a domain for its pose using constraints from camera measurements and distance to the base station. Then, in a second part, a cooperative localization method is introduced, in which neighbours positions and distances are

I.-F. Kenmogne · V. Drevelle (✉) · E. Marchand
Univ Rennes, Inria, CNRS, IRISA, Rennes, France
e-mail: vincent.drevelle@irisa.fr

E. Marchand
e-mail: eric.marchand@irisa.fr

© Springer Nature Switzerland AG 2020
M. Ceberio and V. Kreinovich (eds.), *Decision Making under Constraints*,
Studies in Systems, Decision and Control 276,
https://doi.org/10.1007/978-3-030-40814-5_16

Fig. 1 Cooperative
localization with camera and
range measurements

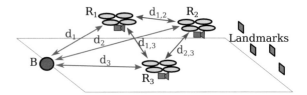

introduced as additional constraints to tighten the pose domain of each robot, thanks
to data exchange. We finally provide experimental results obtained with quadcopter
mini-drones.

2 Vision-Based Pose Computation

This section addresses pose estimation [3] from camera measurements. To compute
the pose $\mathbf{r} = (x, y, z, \phi, \theta, \psi)$ of a UAV amounts to estimating the transformation
$^r\mathbf{T}_w$ between the world reference frame and a reference frame attached to the robot.
We assume that the rigid transformation $^c\mathbf{T}_r$ between the camera and the robot frame
is known from calibration [4], and the camera is calibrated.

For a known landmark of 3D coordinates $^w\mathbf{X}$ in the world reference frame, the
normalized coordinates $\mathbf{x} = (^c x, ^c y)$ of its projection in the camera frame are given
by the pinhole model [5]:

$$\mathbf{x} = \Pi \, ^c\mathbf{T}_r \, ^r\mathbf{T}_w(\mathbf{r}) \, ^w\mathbf{X} \tag{1}$$

where Π is the perspective projection operator.

For each visible landmark $^w\mathbf{X}_i$ ($i \in 1\ldots m$), we can derive the following constraints:

$$C_i : \begin{cases} (^cX_i, ^cY_i, ^cZ_i) = \, ^c\mathbf{T}_r \, ^r\mathbf{T}_w(\mathbf{r}) \, ^w\mathbf{X}_i \\ ^c x_i = \frac{^cX_i}{^cZ_i}, \\ ^c y_i = \frac{^cY_i}{^cZ_i}, \\ ^c x_i \in [^c x_i], \; ^c y_i \in [^c y_i] \quad \text{bounded error camera measurement} \\ ^cZ_i > 0 \quad\quad\quad\quad\quad\quad \text{front looking camera} \end{cases} \tag{2}$$

We then define the image-based pose estimation problem as a constraint satisfaction problem (CSP) as:

$$\mathcal{H}_{img} : \begin{pmatrix} \mathbf{r} \in [\mathbf{r}], \\ \{C_i, \; i \in 1\ldots m\} \end{pmatrix} \tag{3}$$

This CSP is solved with Contractor Programming [6] and Set Inversion via Interval
Analysis (SIVIA) [7]. The altitude $[z]$, pitch $[\theta]$ and roll $[\phi]$ components of the initial
domain $[\mathbf{r}]$ are set thanks to onboard sensors (altimeter and inertial measurement
unit). An outer approximation of the feasible domain for the pose $\mathbf{r}$ is obtained in

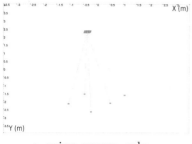

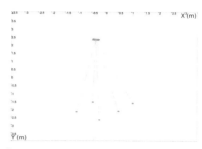

a. using camera only   b. using camera and range to base

Fig. 2 Pose domain computation for a single drone (horizontal projection)

the form of a subpaving. An example of solution set for one robot is shown in Fig. 2a (see [8]).

Assuming a bounded error measurement $[d]$ of the distance d between the robot and the base station B is available, it can be used as an additional constraint on the robot position $\mathbf{p} = (x, y, z)$, to get a tighter pose estimate:

$$d = \|\mathbf{p} - \mathbf{b}\|_2, \quad d \in [d] \tag{4}$$

where $\mathbf{b}$ the known position of the base station. We obtain a tighter domain as presented in Fig. 2b.

Dealing with camera tracking failures The subpaving computed with SIVIA can be empty. This situation corresponds to inconsistencies in the measurements, due to failure of the visual tracking of one or several landmarks (see [8] for more details). In this case, our approach discards the image measurements and use only the distance to the base station B to compute $\mathbb{S}_{\mathbf{r}_k}^+$. The solution-set is in this case a ring centered on B.

3 Using Range Measurements for Cooperative Localization

Assuming each UAV is equipped with a communication system with ranging capabilities, each robot therefore measures ranges to its neighbours and cooperates with them by exchanging its position.

Sharing positions Lets consider one robot R_k ($k \in 1 \dots N$) from the group. $\mathcal{N}(k)$ denotes the neighbours of R_k, i.e. the robots within communication range.

At each time step, R_k first computes an outer subpaving $\mathbb{S}_{\mathbf{r}_k}^+$, that contains all the feasible poses, considering the camera and base distance bounded-error measurements (as presented in Sect. 2).

Once the pose domain $\mathbb{S}_{\mathbf{r}_k}^+$ is computed, the robot computes the bounding box of its position domain $[\mathbf{p}_k] = \Box\,\mathrm{proj}_{\mathbf{p}}\,\mathbb{S}_{\mathbf{r}_k}^+$, where $\Box$ is the bounding box operator, and

$proj_\mathbf{p}$ is the projection on the position space. $[\mathbf{p}_k]$ is transmitted to all neighboring robots R_j, $j \in \mathcal{N}(k)$, and the distances $d_{k,j}$ between R_k and R_j are simultaneously measured.

Pose contraction At reception of information (position boxes $[\mathbf{p}_j]$ and bounded-error distances measurements $[d_{k,j}]$) from neighboring robots, each robot R_k tries to refine its actual pose domain, by propagating the new distance constraints between R_k and each of its neighbours. Recalling that $\mathbf{p}_k = (x_k, y_k, z_k)$ is the position of R_k, each robot R_k contracts a local CSP $\mathcal{H}_k$ defined as follows:

$$
\mathcal{H}_k : \begin{pmatrix} \mathbf{p}_k \in proj_\mathbf{p}(\mathbb{S}^+_{\mathbf{r}_k}), \\ \mathbf{p}_j \in [\mathbf{p}_j], \; j \in \mathcal{N}(k) \\ d_{k,j} \in [d_{k,j}], \; j \in \mathcal{N}(k) \\ d_{k,j} = \|\mathbf{p}_k - \mathbf{p}_j\|_2, \; j \in \mathcal{N}(k) \end{pmatrix} \tag{5}
$$

We use interval constraint propagation to solve $\mathcal{H}_k$, in order to reduce the pose domain $\mathbb{S}^+_{\mathbf{r}_k}$.

Fixed-point The constraint network formed by the group of robots contains cycles spanning several robots. The contraction of the local CSPs $\mathcal{H}_k$ has to be propagated again through the network to improve the pose domain reduction of each robot. If after solving $\mathcal{H}_k$, the robot position bounding box $[\mathbf{p}_k]$ is reduced, then the robot R_k retransmits its updated $[\mathbf{p}_k]$ to its neighbourhood. This process is iterated until a fixed-point is reached (no more significant improvement of the robots positions bounding boxes).

4   Experimental Results

The proposed method has been tested with data acquired on Parrot AR-Drone2 UAV, with 5 landmarks made with AprilTag markers (Fig. 3). Image measurement error bounds are set to ± 0.5 px and range measurement error is assumed to be within ± 5 cm.

Subpavings obtained with 4 robots in cooperative localization are presented in Fig. 4. Figure 4b clearly shows how cooperative localization reduces the feasible pose domain when some robots cannot clearly see the landmarks, by propagating position information of the neighbours.

Table 1 shows how making more robots cooperate in the fleet improves localization, first by reducing the width of the computed position domain (Table 1a), and also

Fig. 3 Onboard cameras views at $t = 8$ s. Landmarks are boxes with a printed pattern

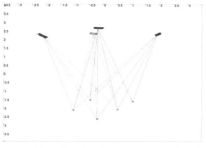

a. $t = 0$ s, the 4 robots cameras track landmarks

b. $t = 6.5$ s, the cameras of R_1 and R_2 are unable to track landmarks

Fig. 4 Position domains of 4 robots. *Black outline: subpavings before communication. Colored domains: subpavings after cooperative localization*

Table 1 Horizontal position results, for 1 to 4 robots in the group

(a)

	R_1	R_2	R_3	R_4
1 UAV	1.29			
2 UAVs	0.86	0.56		
3 UAVs	0.80	0.52	2.98	
4 UAVs	0.33	0.37	0.78	0.61

Horizontal position domain width (m)

(b)

	R_1	R_2	R_3	R_4
1 UAV	0.226			
2 UAVs	0.159	0.088		
3 UAVs	0.141	0.071	1.220	
4 UAVs	0.040	0.046	0.248	0.192

Average horizontal position error (m)

by improving the precision when using the center of the domain as a point estimate (Table 1b). With 4 UAVs, the average horizontal position error is less than 5 cm for all the drones.

5 Conclusion

In this paper, we proposed a method to solve cooperative localization in a group of UAVs. Computations rely on interval constraint propagation, assuming bounded error image and distance measurements. Each UAV first independently computes a pose domain from its camera measurements, and then exchanges position information with its neighbours to further reduce the position domains of the robots in the group. The method has been applied to real data, and enables to improve the positioning precision of the AUVs thanks to position information propagation. The experiments show that increasing the number of robots in the group provides additional constraints on position, and yields smaller uncertainty of the computed robots poses.

References

1. Moore, R.E.: Interval Analysis. Prentice Hall (1966)
2. Roumeliotis, S., Bekey, G., et al.: Distributed multirobot localization. IEEE Trans. Robot. Autom. **18**(5), 781–795 (2002)
3. Marchand, E., Uchiyama, H., Spindler, F.: Pose estimation for augmented reality: a hands-on survey. IEEE Trans Vis. Comput. Graph. **22**(12), 2633–2651 (2016)
4. Tsai, R.: A versatile camera calibration technique for high-accuracy 3D machine vision metrology using off-the-shelf tv cameras and lenses. IEEE J. Robot. Autom. **3**(4), 323–344 (1987)
5. Hartley, R., Zisserman, A.: Multiple View Geometry in Computer Vision, 2nd edn. Cambridge University Press (2004)
6. Chabert, G., Jaulin, L.: Contractor programming. Artif. Intell. **173**(11), 1079–1100 (2009)
7. Jaulin, J., Walter, E.: Set inversion via interval analysis for nonlinear bounded-error estimation. Automatica **29**(4), 1053–1064 (1993)
8. Kenmogne, I.F., Drevelle, V., Marchand, E.: Image-based UAV localization using interval methods. In: IEEE/RSJ International Conference on Intelligent Robots and Systems, Sept 2017, pp. 5285–5291

Attraction-Repulsion Forces Between Biological Cells: A Theoretical Explanation of Empirical Formulas

Olga Kosheleva, Martine Ceberio and Vladik Kreinovich

Abstract Biological cells attract and repulse each other: if they get too close to each other, they repulse, and if they get too far away from each other, they attract. There are empirical formulas that describe the dependence of the corresponding forces on the distance between the cells. In this paper, we provide a theoretical explanation for these empirical formulas.

1 Formulation of the Problem

Biological calls interact. Biological cells attract and repulse each other. For each type of cell, there is a certain distance R_1 at which there is no interaction.

- When the cells get closer to each other than this threshold distance, i.e., when the distance r between the cells becomes smaller than R_1, then the cells repulse each other.
- On the other hand, if the two cells deviate further away that the threshold distance R_1, they start attracting each other.

As a result of these two forces, the cells stay at the same—biologically optimal—distance from each other.

Empirical formulas describing interaction between the cells. According to [2, 3, 5], the interaction force $\mathbf{f}$ between the two cells at a distance $\mathbf{r}$ has the following form:

O. Kosheleva · M. Ceberio · V. Kreinovich (✉)
University of Texas at El Paso, El Paso, TX 79968, USA
e-mail: vladik@utep.edu

O. Kosheleva
e-mail: olgak@utep.edu

M. Ceberio
e-mail: mceberio@utep.edu

© Springer Nature Switzerland AG 2020
M. Ceberio and V. Kreinovich (eds.), *Decision Making under Constraints*,
Studies in Systems, Decision and Control 276,
https://doi.org/10.1007/978-3-030-40814-5_17

- when $r < R_1$, we have $\mathbf{f} = -k_1 \cdot \left(\dfrac{1}{r} - \dfrac{1}{R_1} \right) \cdot \mathbf{e}$, where r is the length of the vector $\mathbf{r}$ (i.e., the distance between the cells) and $\mathbf{e} \overset{\text{def}}{=} \dfrac{\mathbf{r}}{r}$ is the unit vector in the direction $\mathbf{r}$;
- when $r > R_1$, we have $\mathbf{f} = k_2 \cdot (r - R_1) \cdot \mathbf{e}$.

Formulation of the problem. How can we explain these empirical formulas?

What we do in this paper. In this paper, we provide a theoretical explanation for the above empirical formulas.

2   Analysis of the Problem

Qualitative requirements. We want to find the dependence $f(r)$ of the interactive force f on the distance r between the two cells.

To find such a dependence, let us consider natural requirements on $f(r)$.

Monotonicity. The larger the difference between the actual distance r and the desired distance R_1, the larger should be the force. Thus:

- the repulsion force should increase when the distance r decreases, while
- the attraction force should increase as the distance r increases.

It should be mentioned that the empirical formulas satisfy this property—i.e., the corresponding dependencies $f(r)$ are monotonic.

Analyticity. This is a general phenomenon in physics, that all dependencies in fundamental physics are described by analytical functions, i.e., by functions which can be expanded in Taylor or, more generally, by Laurent series; see, e.g., [4]. For functions of one variable, this means that we must have

$$f(r) = a_0 + a_1 \cdot r + a_2 \cdot r^2 + \cdots + a_{-1} \cdot r^{-1} + a_{-2} \cdot r^{-2} + \cdots$$

In fundamental physics, this phenomenon is usually explained by the need to consider quantum effects: quantum analysis means extension to complex numbers – and analytical functions are, in effect, differential functions of complex variables; see, e.g., [4].

Again, it is worth mentioning that both empirical formulas—the formula corresponding to $r < R_1$ and the formula corresponding to $r > R_1$—are analytical.

Tidal forces. The main objective of the forces between the two cells are to keep the cells at a certain distance. This motivates the direct effect of the forces:

- when the cells are two close, the repulsion force will make move apart, while
- when the cells are too far away from each other, the attraction force will make them get closer.

However, with this direct effect, there is also an undesired side effect, caused by the fact that cells are not points. As a result, different parts of the cell have slightly difference force acting on them. So, in addition to the overall force that makes the cell move in the desired direction, we also have tidal forces that make the parts of the cell move with respects to each other—i.e., make the cell compress or stretch.

In general, the tidal forces are proportional to the gradient of the force field (see, e.g., [4]), i.e., in this case, to the derivative $F(r) \stackrel{\text{def}}{=} \dfrac{df}{dr}$.

From the biological viewpoint, tidal forces are undesirable, so they should be as small as possible.

Scale invariance. Physical laws are formulated in terms of the numerical values of physical quantities. However, these numerical values depend on what measuring unit we select to describe this quantity.

For example, if we first measure distances in meters, and then start using centimeters instead, then all the numerical values get multiplied by a factor of 100. In particular, 2 m becomes 200 cm.

In most fundamental physical laws, there is no physically preferred unit. In such situations, it make sense to require that the physical law not depend on the choice of the unit.

Of course it does not means that all the formulas remain unchanged if we simply change the measuring unit of one of the quantities. Usually, if we change the unit of one of the quantities, then we have to accordingly change the units of related quantities. However, after an appropriate re-scaling of all the units, all the formulas should remain the same.

In precise terms, scale-invariance of the dependence $b = B(a)$ between two quantities a and b means that for every λ, there exists a $\mu(\lambda)$ such that if we change a to $a' = \lambda \cdot a$ and b to $b' = \mu(\lambda) \cdot b$, the dependence remain the same: if $b = B(a)$, then we should have $b' = B(a')$, i.e., $\mu(\lambda) \cdot b = B(\lambda \cdot a)$.

For the dependence of the force itself of the distance, there is clearly no scale-invariance: indeed, in this case, there is a special distance R_1 at which the force is 0. However, for the *tidal force* $F(r)$, interestingly, there is scale-invariance: namely, $F(r) \sim r^{-2}$ for small r and $F(r) = \text{const}$ for large r; both are scale-invariant formulas.

Now, we are ready to describe our result.

3 Definitions and the Main Result

The above properties take the following form:

Definition 1

- By a *force function*, we mean a function $f(r)$ from positive numbers to real numbers.

- We say that a force function unction $f(r)$ is *analyticaL* if it can be expanded in Laurent series

$$f(r) = a_0 + a_1 \cdot r + a_2 \cdot r^2 + \cdots + a_{-1} \cdot r^{-1} + a_{-2} \cdot r^{-2} + \cdots$$

- We say that a force function is *monotonic-at-0* if for sufficiently small r, its absolute value increases as r decreases.
- We say that a force function is *monotonic-at-∞* if for sufficiently large r, its absolute value increases as r increases.
- By a *tidal force function* corresponding to the force function $f(r)$, we mean its derivative $F(r) = \dfrac{df}{dr}$.
- We say that a tidal force function is *scale-invariant* if for every $\lambda > 0$, there exists a $\mu(\lambda)$ for which, for all r and a, $a = F(r)$ implies that $\mu(\lambda) \cdot a = F(\lambda \cdot r)$.

Next, we should describe the property that the tidal forces should not grow too fast.

Definition 2 Let $f(r)$ be an analytical monotonic-at-0 force function $f(r)$, let $F(r)$ be its tidal force function, and let $F(r)$ be scale-invariant. We say that $F(r)$ *grows fast* if there exists another analytical monotonic-at-0 force function $g(r)$, with scale-invariant tidal force function $G(r)$, for which $\dfrac{F(r)}{G(r)} \to \infty$ as $r \to 0$.

Definition 3 Let $f(r)$ be an analytical monotonic-at-∞ force function $f(r)$, let $F(r)$ be its tidal force function, and let $F(r)$ be scale-invariant. We say that $F(r)$ *grows fast* if there exists another analytical monotonic-at-∞ force function $g(r)$, with scale-invariant tidal force function $G(r)$, for which $\dfrac{F(r)}{G(r)} \to \infty$ as $r \to 0$.

Proposition 1 *Every analytical monotonic-at-0 force function $f(r)$ for which the tidal force function $F(r)$ is scale-invariant and does not grow fast, has the form* $f(r) = \dfrac{c_0}{r} + c_1$ *for some c_0 and c_1.*

Proposition 2 *Every analytical monotonic-at-∞ force function $f(r)$ for which the tidal force function $F(r)$ is scale-invariant and does not grow fast, has the form* $f(r) = c_0 \cdot r + c_1$ *for some c_0 and c_1.*

Discussion. These are exactly the empirical formulas that we wanted to explain. Thus, we have a theoretical explanation for these formulas.

4 Proofs

$1°$. Let us first see what we can conclude from scale-invariance of the tidal force function $F(r)$. By definition, this scale-invariance means that

$$F(\lambda \cdot r) = \mu(\lambda) \cdot F(r).$$

The function $F(r)$ is analytical, thus smooth. For smooth functions, every function $F(r)$ with this property has the form $F(r) = c \cdot r^\alpha$ for some constants c and α; see, e.g., [1].

This fact is easy to prove. Since the function $F(r)$ is smooth, the function $\mu(\lambda)$, which is equal to the ratio of two smooth functions $\mu(\lambda) = \dfrac{F(\lambda \cdot r)}{F(r)}$, is also smooth. Differentiating both sides of the equality $F(\lambda \cdot r) = \mu(\lambda) \cdot F(r)$ by λ and taking $\lambda = 1$, we conclude that $r \cdot \dfrac{dF}{dr} = \alpha \cdot F$, where $\alpha \stackrel{\text{def}}{=} \dfrac{d\mu}{d\lambda}_{|\lambda=1}$.

By moving all the terms containing r to one side and all the terms containing F to another side, we conclude that $\dfrac{dF}{F} = \alpha \cdot \dfrac{dr}{r}$. Integrating both sides, we get $\ln(F(r)) = \alpha \cdot \ln(r) + C$, for some integration constant C. Thus, for $F(r) = \exp(\ln(F(r)))$, we get $F(r) = c \cdot r^\alpha$, where $c \stackrel{\text{def}}{=} \exp(C)$.

$2°$. Since the force function $f(r)$ is analytical, its derivative is also analytical. Thus, α should be an integer.

For $\alpha = -1$, integration of the above expression for $F(r)$ would lead to $f(r) = c \cdot \ln(r)$, which is not an analytical function. Thus, $\alpha \neq -1$, and integration of $F(r)$ leads to $f(r) = c_0 \cdot r^{\alpha+1} + c_1$, where $c_0 \stackrel{\text{def}}{=} \dfrac{c}{\alpha + 1}$.

$3°$. Monotonicity-at-0 implies that $\alpha + 1 < 0$, i.e., that that $\alpha + 1 \leq -1$ and $\alpha \leq -2$. For $\alpha < -2$, we could take $g(r) = r^{-1}$ with $G(r) = -r^{-2}$ and thus,

$$\frac{F(r)}{G(r)} \sim \frac{r^\alpha}{r^{-2}} = r^{\alpha+2}.$$

From $\alpha < -2$, it follows that $\alpha + 2 < 0$, hence $\dfrac{F(r)}{G(r)} \sim r^{\alpha+2} \to \infty$ as $r \to 0$. So, all the cases when $\alpha < -2$ correspond to the tidal force function that grows fast. The only case when this function does not grow fast is the case $\alpha = -2$, which leads to

$$f(r) = c_0 \cdot r^{-1} + c_1.$$

$4°$. Similarly, monotonicity-at-∞ implies that $\alpha + 1 > 0$, i.e., that $\alpha + 1 \geq 1$ and therefore $\alpha \geq 0$.

For $\alpha > 0$, we could take $g(r) = r$ with $G(r) = 1$ and thus, $\dfrac{F(r)}{G(r)} \sim r^\alpha$. From $\alpha > 0$, $r^\alpha \to \infty$ as $r \to \infty$. So, all the cases when $\alpha > 0$ correspond to the tidal force function that grows fast. The only case when this function does not grow fast is the case $\alpha = 0$, which leads to $f(r) = c_0 \cdot r + c_1$.

The Propositions are proven.

Acknowledgements This work was supported in part by NSF grants HRD-0734825, HRD-1242122, and DUE-0926721, and by an award from Prudential Foundation. The authors are thankful to Dr. Robert Smits from New Mexico State University for valuable discussions.

References

1. Aczél, J., Dhombres, H.: Functional Equations in Several Variables. Cambridge University Press, Cambridge, UK (1989)
2. Di Costanzo, E., Giacomello, A., Messina, E., Natalini, R., Pontrelli, G., Rossi, F., Smits, R., Twarogowska, M.: A discrete in continuous mathematical model of cardiac progenitor cells formation and growth as spheroid clusters (cardiospheres). Math. Med. Biol (2017)
3. D'Orsogna, M.R., Chuang, Y.L., Bertozzi, A.L., Chayes, L.S.: Self-propelled particles with soft-core interactions: patterns, stability, and collapse. Phys. Rev. Lett. **96**, 104302–104305 (2016)
4. Feynman, R.P., Leighton, R.B., Sands, M.L.: The Feynman Lectures on Physics. Addison-Wesley, Redwood City, California (2005)
5. Joie, J., Lei, Y., Durrieu, M.-C., Colin, T., Poignard, C., Saut, O.: Migration and orientation of endothelial cells on micropatterned polymers: a simple model based on classical mechanics. Discret. Contin. Dyn. Syst. Ser. B **20**, 1059–1076 (2015)

When We Know the Number of Local Maxima, Then We Can Compute All of Them

Olga Kosheleva, Martine Ceberio and Vladik Kreinovich

Abstract In many practical situations, we need to compute local maxima. In general, it is not algorithmically possible, given a computable function, to compute the locations of all its local maxima. We show, however, that if we know the *number* of local maxima, then such an algorithm is already possible. Interestingly, for global maxima, the situation is different: even if we only know the number of locations where the *global* maximum is attained, then, in general, it is not algorithmically possible to find all these locations. A similar impossibility result holds for local maxima if instead of knowing their exact number, we only know two possible numbers.

1 Locating Local Maxima: An Important Practical Problem

Need for computing local maxima. In many practical situations, we are interested in locating all local optima; see, e.g., [3]. For example:

- in spectral analysis, chemical species are identified by local maxima of the spectrum;
- in radioastronomy, radiosources and their components are identified as local maxima of the brightness distribution; see, e.g., [4];
- elementary particles are identified by locating local maxima of the dependence of scattering intensity on the energy.

O. Kosheleva · M. Ceberio · V. Kreinovich (✉)
University of Texas at El Paso, El Paso, TX 79968, USA
e-mail: vladik@utep.edu

O. Kosheleva
e-mail: olgak@utep.edu

M. Ceberio
e-mail: mceberio@utep.edu

© Springer Nature Switzerland AG 2020
M. Ceberio and V. Kreinovich (eds.), *Decision Making under Constraints*,
Studies in Systems, Decision and Control 276,
https://doi.org/10.1007/978-3-030-40814-5_18

In general, no algorithm is possible for computing all local maxima. In general, no algorithm is possible for computing all local maxima of a computable function $f(x)$; this follows, e.g., from our negative result formulated below.

Natural question and what we do in this paper. Since we cannot *always* compute all local maxima, a natural question is: *when* can we compute them? In this paper, we prove that such a computation is algorithmically possible in situations when we know the number of local maxima—and not possible if we only know two possible candidates for this number.

2   What Is Computable: Reminder

Need for a reminder. To formulate our results, we need to recall the main definitions of what is computable: what is a computable number, what is a computable set, and what is a computable function; see, e.g., [1, 2, 5] for more details.

What is a computable number: intuitive idea. In a computer, we can only represent rational numbers—namely, binary-rational ones. Thus, it is reasonable to say that a real number is computable if it can be algorithmically approximated, with any given accuracy, by rational numbers.

What is a computable number: a precise definition. A real number x is called *computable* if there is an algorithm that, given a natural number n, produces a rational number r_n which is 2^{-n}-close to x.

What is a computable set: intuitive idea. In a computer, we can only store finitely many objects—i.e., a finite set, with computable distances. It is therefore reasonable to define a computable set as a set that can be algorithmically approximated, with any given accuracy, by finite sets—approximated in the sense that every element of our set is 2^{-n}-close to one of the elements from the approximating finite set.

Since a computer has a linear memory, it is convenient to place all the elements of these finite sets—which approximate our set with higher and higher accuracy—into a single infinite sequence x_1, x_2, ... Elements from this sequence approximate any element from the given set. Thus, this sequence must be *everywhere* dense in this set.

In practice, we do not know the exact values of the elements, we only have approximations to elements of the set. Based on these approximations, we can never know whether the resulting set is closed or not—i.e., whether a set of real numbers is the interval $[-1, 1]$ or the same interval minus 0 point. To ignore such un-detectable differences, it is reasonable to assume that our set is *complete*, i.e., that it includes the limit of each converging sequence.

Thus, we arrive at the following definition.

What is a computable set: definition. By a *computable set*, we mean a complete metric space with an everywhere dense sequence $\{x_i\}$ for which:

- there is an algorithm that, given i and j, computes the distance $d(x_i, x_j)$ (with any given accuracy), and
- there exists an algorithm that, given a natural number n, returns a natural number $N(n)$ for which every point $x_1, x_2, \ldots$ is 2^{-n}-close to one of the points

$$x_1, \ldots, x_{N(n)}.$$

By a *computable element* x of a computable set, we mean an algorithm that, given a natural number n, returns an integer $i(n)$ for which $d(x, x_{i(n)}) \leq 2^{-n}$.

Comment. From the topological viewpoint, a complete metric space which can be approximated by finite sets is a *compact space*. Thus, computable sets are also known as *computable compact sets*.

What is a computable function: intuitive idea. A computable function f should be able, given a computable real number (or, more generally, a computable element of a computable set), to compute the value $f(x)$ with any given accuracy. Computable elements x are given by their approximations. Thus, to compute $f(x)$ with a given accuracy 2^{-n}, we need to:

- first algorithmically determine how accurately we need to compute x to achieve the desired accuracy 2^{-n} in $f(x)$, and then
- use the corresponding approximation to x to actually compute the desired approximation to $f(x)$.

So, we arrive at the following definition.

What is a computable function: definition. We say that a function $f(x)$ from a computable set to real numbers is computable if:

- first, we have an algorithm that, given n, returns m for which $d(x, x') \leq 2^{-m}$ implies that $|f(x) - f(x')| \leq 2^{-n}$, and
- second, we have an algorithm that, given i, computes $f(x_i)$.

Comment. The existence of m for every n is nothing else but uniform continuity; so, in effect, we want $f(x)$ to be effectively uniformly continuous.

Now, we are ready to formulate our main results.

3 Main Results

Proposition 1 *There exists an algorithm that:*

- *given an integer m and a computable function $f(x)$ with exactly m local maxima,*
- *always computes the locations of all these maxima.*

Comment 1. It is worth mentioning that for *global* maxima, such an algorithm is not possible even for $m = 2$: no algorithm can, given a computable function at which the global maximum is attained at exactly two points, computes these two locations; see, e.g., [2].

Comment 2. Knowing the exact number of local maxima is important: as the following result shows, if we have an *incomplete* information about this number, we can no longer compute all the local maxima.

Proposition 2 *Let $m < m'$ be two natural numbers. Then, no algorithm is possible that:*

- *given a computable function $f(x)$ with either m or m' local maxima,*
- *always returns the locations of all its local maxima.*

4   Proof of Proposition 1

Auxiliary results needed to describe our algorithm. Our algorithm is based on the several known results. The first is that we can algorithmically compute the maximum of a computable function on a computable set; see, e.g., [1, 2, 5].

We will also use an easy-to-prove fact that for every computable element x_0, the function $f(x) \overset{\text{def}}{=} d(x, x_0)$ is computable; see, e.g., [1, 5].

Another result that we will use is that for every computable function $f(x)$ on a computable set, and for every four rational numbers $\underline{r}_1 < \overline{r}_1 < \underline{r}_2 < \overline{r}_2$, we can algorithmically find the values $b_1 \in (\underline{r}_1, \overline{r}_1)$ and $b_2 \in (\underline{b}_2, \overline{b}_2)$ for which the set

$$\{x : b_1 \le f(x) \le b_2\}$$

is also a computable set; see, e.g., [1].

We will also use the fact that each positive rational number p/q is simply a pair of natural numbers. Thus, a tuple consisting of natural and positive rational numbers can be viewed simply as a tuple consisting of natural numbers.

We can algorithmically sort the tuples consisting of positive natural numbers: e.g., first we consider all (finitely many) tuples whose sum is 1, then all the tuples whose sum is 2, etc.

Now, we are ready to describe our algorithm.

Our algorithm: a description. We want to locate all the maxima with a given accuracy 2^{-n}. To do that, by using one of above-mentioned sorting, we try, one by one, all possible tuples consisting of two natural numbers i and k and four positive rational numbers for which $\underline{r}_1 < \overline{r}_1 < \underline{r}_2 < \overline{r}_2 \le 2^{-n}$. For each such tuple, we compute $f \overset{\text{def}}{=} f(x_i)$ and $s \overset{\text{def}}{=} \max\{f(x) : b_1 \le d(x, x_i) \le b_2\}$ (for appropriate $b_i \in (\underline{r}_i, \overline{r}_i)$) with accuracy 2^{-k}. If for the resulting approximations $\widetilde{f}$ and $\widetilde{s}$, we get $\widetilde{f} > \widetilde{s} + 2 \cdot 2^{-k}$, then we can conclude that $f > s$.

We stop when we get m different tuples such that:

- each of these m tuples satisfies the inequality $\widetilde{f} > \widetilde{s} + 2 \cdot 2^{-k}$ (thus $f > s$), and

- for every two tuples, the distance $d(x_i, x_j)$ between the corresponding elements x_i and x_i' is larger than the sum of the corresponding values $\bar{r}_2$ and $\bar{r}_2'$.

The corresponding m elements x_i are then returned as the desired 2^{-n}-approximations to the locations of the local maxima.

What we need to prove. To prove the proposition, we need to prove:

- that this algorithm always stops, and that
- that this algorithm is correct, i.e., that the results of this algorithm are indeed 2^{-n}-approximations to the desired locations of local maxima.

Let us first prove that the algorithm always stops. Let $M_1, \ldots, M_m$ be the desired local maxima, and let d_0 be the smallest of all the distances between them.

By definition, a local maximum means that $f(M_j) \geq f(x)$ for all x from some neighborhood of M_j. We can always select this neighborhood of size $\leq d_0/3$. This way, we can be sure that there are no other local maxima in this neighborhood—and thus, no values $x \neq M_j$ with $f(M_j) = f(x)$, since otherwise these values x will also be local maxima.

Therefore, $f(M_j) > f(x)$ for all all the values from this neighborhood.

Let δ be a rational number which is smaller than all m radii of these neighborhoods. Then $f(M_j) > f(x)$ for all x for which $\delta/2 \leq d(x, M_j) \leq \delta$. The set $\{x : \delta/2 \leq d(x, M_j) \leq \delta\}$ is a compact, so for a continuous function $f(x)$, the maximum is attained at some element from this set. Since for all the points from this set, we have $f(x) < f(M_j)$, we therefore conclude that

$$f(M_j) > \max\{f(x) : \delta/2 \leq d(x, M_j) \leq \delta\}.$$

Due to continuity, for elements x_i which are sufficiently close to M_j, we also have

$$f(x_i) > \max\{f(x) : \delta/2 \leq d(x, M_j) \leq \delta\}.$$

Here, if $d(x_i, M_j) \leq \varepsilon$, then $\delta/2 + \varepsilon \leq d(x, x_i) \leq \delta - \varepsilon$ imply $\delta/2 \leq d(x, M_j) \leq \delta$. Thus:

$$\max\{f(x) : \delta/2 \leq d(x, M_j) \leq \delta\} \geq \max\{f(x) : \delta/2 + \varepsilon < d(x, x_i) \leq \delta - \varepsilon\}$$

and

$$f(x_i) > \max\{f(x) : \delta/2 + \varepsilon < d(x, x_i) \leq \delta - \varepsilon\}.$$

So, when we take $\underline{r}_1 = \delta/2 + \varepsilon$ and $\bar{r}_2 = \delta - \varepsilon$, we will get

$$f(x_i) > \max\{f(x) : b_1 < d(x, x_i) \leq b_2\}.$$

Whenever the above strict inequality is true, we will detect it if we compute both sides of this inequality with sufficient accuracy. Thus, eventually, we will indeed find the tuples for which $\tilde{f} > \tilde{s} + 2 \cdot 2^{-k}$ and for which each x_i is desirably close to the corresponding local maximum M_j. Hence, our algorithm will indeed always stop.

Let us now prove that the algorithm is correct. To complete our proof, we need to show that when the algorithm stops, the resulting elements x_i are indeed close to the corresponding local maxima. Indeed, when the algorithm stops, for each of the selected m tuples, we get $f(x_i) > \max\{f(x) : b_1 \leq d(x, x_i) \leq b_2\}$.

On the compact set $\{x : d(x, x_i) \leq b_2\}$, the maximum of the continuous function $f(x)$ is attained at some element from this set. Due to the above inequality, this maximum cannot be attained at distances between b_1 and b_2. Thus, this maximum is attained when $d(x, x_i) \leq b_1 < b_2$. So, this maximum is a local maximum of the function $f(x)$, and a local maximum which is (due to $b_2 < \bar{r}_2 \leq 2^{-n}$) 2^{-n}-close to the corresponding element x_i.

On each "zone" $\{x : d(x, x_i) \leq b_2\}$ we thus have a local maximum of the given function $f(x)$. Since $d(x_i, x_j) > \bar{r}_2 + \bar{r}_2' > b_2 + b_2'$, these zones do not intersect. Thus:

- all m corresponding local maxima are different,
- there are no local maxima outside these zones, and
- within each zone, we have exactly one local maximum which is 2^{-n}-close to x_i.

The correctness is proven, and so is the proposition.

5   Proof of Proposition 2

Our proof-by-contradiction is based on the fact that no algorithm is possible that, given a non-negative real number a, checks whether $a = 0$. This result can be easily proven based on the halting problem result. Indeed, for each Turing machine, we can define a computable real number a for which:

- $r_n = 2^{-n}$ if this Turing machine did not halt by moment n and
- $r_n = 2^{-t}$ if it halted at moment $t \leq n$.

As a result:

- If the Turing machine does not halt, then the resulting number is equal to 0.
- Otherwise, if the Turing machine halts at some time t, then we have $a = 2^{-t} > 0$.

The impossibility to check whether a Turing machine halts implies that we cannot check whether $a = 0$.

For each m and m', let us define the following function $f(x)$ on the interval

$$[0, 2m + 2(m' - m) \cdot a] :$$

For $x \in [0, 2m]$, we have:

- $f(x) = 1 - |x - 1|$ for $0 \leq x \leq 2$,
- $f(x) = 1 - |x - 3|$ for $2 \leq x \leq 4$,
- ...,

- $f(x) = 1 - |x - (2m - 1)|$ when $2m - 2 \leq x \leq 2m$.

For $x \in [2m, 2m + 2(m' - m) \cdot a]$, we have:

- $f(x) = a - |x - (2m + a)|$ when $2m \leq x \leq 2m + 2a$,
- $f(x) = a - |x - (2m + 3a)|$ when $2m + 2a \leq x \leq 2m + 4a$,
- ...,
- $f(x) = a - |x - (2m + (2(m' - m) - 1) \cdot a)|$ when

$$2m + (2(m' - m) - 2) \cdot a \leq x \leq 2m + 2(m - m) \cdot a.$$

Here:

- When $a = 0$, this function has m local maxima, at points $1, 3, \ldots, 2m - 1$.
- When $a > 0$, this function has $m + (m' - m) = m'$ local maxima:

  - m local maxima at points $1, 3, \ldots, 2m - 1$, and
  - $m' - m$ local maxima at points $2m + a, 2m + 3a, \ldots, 2m + (2(m' - m) - 1) \cdot a$.

If we could always return all the local maxima, then by checking whether there is a local maximum close to $2m$, we would be able to check whether $a > 0$ or $a = 0$, and we have already shown that this is not possible. This proves Proposition 2.

Acknowledgements This work was supported in part by NSF grants HRD-0734825, HRD-1242122, and DUE-0926721, and by an award from Prudential Foundation.

References

1. Bishop, E.: Foundations of Constructive Analysis. McGraw-Hill, New York (1967)
2. Kreinovich, V., Lakeyev, A., Rohn, J., Kahl, P.: Computational Complexity and Feasibility of Data Processing and Interval Computations. Kluwer, Dordrecht (1997)
3. Villaverde, K., Kreinovich, V.: A linear-time algorithm that locates local extrema of a function of one variable from interval measurement results. Interval Comput. **4**, 176–194 (1993)
4. Verschuur, G.L., Kellermann, K.I.: Galactic and Extra-Galactic Radio Astronomy. Springer, Berlin, Heidelberg, New York (1974)
5. Weihrauch, K.: Computable Analysis. Springer, Berlin (2000)

Why Decimal System and Binary System Are the Most Widely Used: A Possible Explanation

Gerardo Muela and Olga Kosheleva

Abstract What is so special about numbers 10 and 2 that decimal and binary systems are the most widely used? One interesting fact about 10 is that when we start with a unit interval and we want to construct an interval of half width, then this width is exactly 5/10; when we want to find a square of half area, its sides are almost exactly 7/10, and when we want to construct a cube of half volume its sides are almost exactly 8/10. In this paper, we show that 2, 4, and 10 are the only numbers with this property—at least among the first billion numbers. This may be a possible explanation of why decimal and binary systems are the most widely used.

1 Formulation of the Problem

Problem. What is so special about numbers 10 and 2 that decimal and binary systems are the most widely used?

This questions was raised, e.g., in [1].

Observation. One interesting fact about 10 is the following:

- When we start with a unit interval and we want to constrict an interval of half width, then this width is exactly $1/2 = 5/10$.
- When we start with a unit square and want to find a square of area 1/2, its sides are $\sqrt{1/2}$, which is almost exactly 7/10:

$$\left| \sqrt{\frac{1}{2}} - \frac{7}{10} \right| < \frac{1}{100}.$$

G. Muela · O. Kosheleva (✉)
University of Texas at El Paso, El Paso, TX 79968, USA
e-mail: olgak@utep.edu

G. Muela
e-mail: gdmuela@miners.utep.edu

© Springer Nature Switzerland AG 2020
M. Ceberio and V. Kreinovich (eds.), *Decision Making under Constraints*,
Studies in Systems, Decision and Control 276,
https://doi.org/10.1007/978-3-030-40814-5_19

- When we start with a unit cube and want to find a cube of volume 1/2, its sides are $\sqrt[3]{1/2}$, which is almost exactly 8/10:

$$\left| \sqrt[3]{\frac{1}{2}} - \frac{8}{10} \right| < \frac{1}{100}.$$

So, whether we want to construct a piece of land which is (almost) exactly of half-area, or a piece of gold which is (almost) exactly of half-volume, decimal systems is very convenient.

Are there any other numbers with this property? Maybe here are other bases b with this property, i.e., bases b for which, for appropriate numbers n_1, n_2, and n_3, we have

$$\left| \frac{1}{2} - \frac{n_1}{b} \right| < \frac{1}{b^2}, \quad \left| \sqrt{\frac{1}{2}} - \frac{n_2}{b} \right| < \frac{1}{b^2}, \quad \left| \sqrt[3]{\frac{1}{2}} - \frac{n_3}{b} \right| < \frac{1}{b^2}. \tag{1}$$

What we do in this paper. In this paper, we show that—at least among the first billion numbers b—only the numbers $b = 2$, $b = 4$, and $b = 10$ satisfy this property.

Base 4 is, in effect, the same as the binary system—we just group two binary digits to get one 4-ary digit, just like we get an 8-ary system when we group three binary digits or 16-based system when we group 4 binary digits together.

Thus, the above result may be a good explanation of why decimal and binary systems are the most widely used.

2   Analysis of the Problem

Considering the first condition. Let us first consider the first of the desired inequalities: $\left| \frac{1}{2} - \frac{n_1}{b} \right| < \frac{1}{b^2}$. When the base is even, i.e., when $b = 2k$ for some integer k, then this property is clearly satisfied: indeed, in this case, for $n_1 = k$, we get $\frac{n_1}{b} = \frac{1}{2}$ and thus, $\left| \frac{1}{2} - \frac{k}{b} \right| = 0 < \frac{1}{b^2}$.

On the other hand, if b is odd, i.e., if $b = 2k + 1$ for some natural number $k \geq 1$, then, for $\frac{1}{2} = \frac{k + 0.5}{2k + 1} = \frac{k + 0.5}{b}$, the closest fractions of the type $\frac{n_1}{b}$ are the fractions $\frac{k}{b}$ and $\frac{k + 1}{b}$. For both these fractions, we have

$$\left| \frac{k + 0.5}{2k + 1} - \frac{k}{2k + 1} \right| = \left| \frac{k + 0.5}{2k + 1} - \frac{k + 1}{2k + 1} \right| = \frac{0.5}{2k + 1} = \frac{1}{2 \cdot (2k + 1)} = \frac{1}{2b}.$$

The desired inequality thus takes the form $\dfrac{1}{2b} < \dfrac{1}{b^2}$, which is equivalent to $2b > b^2$ and $2 > b$. However, odd bases start with $b = 3$. So, the first condition cannot be satisfied by odd bases b.

Thus, the first condition is equivalent to requiring that the base b is an even number.

How do we check the second condition. If we check the second condition $\left| \sqrt{\dfrac{1}{2}} - \dfrac{n_2}{b} \right| < \dfrac{1}{b^2}$ literally, then we need to consider all possible values n_2 from 0 to b. However, this can avoided if we multiply both sides of the desired inequality by b and consider the equivalent inequality $\left| b \cdot \sqrt{\dfrac{1}{2}} - n_2 \right| < \dfrac{1}{b}$. In this case, we can easily see that n_2 is the nearest integer to the product $b \cdot \sqrt{\dfrac{1}{2}}$:

$$n_2 = \left[b \cdot \sqrt{\dfrac{1}{2}} \right],$$

where $[x]$ denotes the nearest integer to the real number x. In these terms, the desired inequality takes the form

$$\left| b \cdot \sqrt{\dfrac{1}{2}} - \left[b \cdot \sqrt{\dfrac{1}{2}} \right] \right| < \dfrac{1}{b}. \tag{2}$$

This is the inequality that we will check.

How to check the third condition. Similarly, if we check the third condition $\left| \sqrt[3]{\dfrac{1}{2}} - \dfrac{n_3}{b} \right| < \dfrac{1}{b^2}$ literally, then we need to consider all possible values n_3 from 0 to b. However, this can avoided if we multiply both sides of the desired inequality by b and consider the equivalent inequality $\left| b \cdot \sqrt[3]{\dfrac{1}{2}} - n_3 \right| < \dfrac{1}{b}$. In this case, we can easily see that n_3 is the nearest integer to the product $b \cdot \sqrt[3]{\dfrac{1}{2}}$:

$$n_3 = \left[b \cdot \sqrt[3]{\dfrac{1}{2}} \right],$$

where $[x]$ denotes the nearest integer to the real number x. In these terms, the desired inequality takes the form

$$\left| b \cdot \sqrt[3]{\dfrac{1}{2}} - \left[b \cdot \sqrt[3]{\dfrac{1}{2}} \right] \right| < \dfrac{1}{b}. \tag{3}$$

This is the inequality that we will check.

The checking. For each even number b from 2 to 10^9, we checked whether this number satisfies both conditions (2) and (3). A simple Java program for this checking is given in Sect. 3.

The result of the checking. The result is that among all the bases b from 1 to 10^9, both roots are only well approximated for $b = 2$, $b = 4$, and $b = 10$. Thus, only for these three bases, the desired condition (1) is satisfied.

This may explain why decimal and binary systems are the most frequently used.

Natural conjecture. We have checked all the values b until 10^9. This makes us conjecture that out of *all* possible natural numbers $b \geq 2$, only the numbers 2, 4, 10 satisfy the property (1).

3   Code

```
public static void main(String [] args){
  double value;
  //Loop that iterates from 2 to 10^9
  for(int b = 2; b <= 1000000000; b += 2){
    value = Math.sqrt(0.5) * b;
    //Checks if the square root is well approximated
    if(Math.abs(value - Math.round(value)) < 1. / b){
      value = Math.cbrt(0.5) * b;
      //Checks if the cubic root is well approximated
      if(Math.abs(value - Math.round(value)) < 1. / b){
        System.out.println("Square and cubic roots "
          + "are well approximated in base " + b);
      }
    }
  }
}
```

Reference

1. Knuth, D.E.: The Art of Computer Programming. Addison-Wesley Professional, Boston, MA (2011)

Uncertainty in Boundary Conditions—An Interval Finite Element Approach

Rafi L. Muhanna and Shahrokh Shahi

Abstract In this work, we introduce an interval formulation that accounts for uncertainty in supporting conditions of structural systems. Uncertainty in structural systems has been the focus of a wide range of research. Different models of uncertain parameters have been used. Conventional treatment of uncertainty involves probability theory, in which uncertain parameters are modeled as random variables. Due to specific limitation of probabilistic approaches, such as the need of a prior knowledge on the distributions, lack of complete information, and in addition to their intensive computational cost, the rationale behind their results is under debate. Alternative approaches such as fuzzy sets, evidence theory, and intervals have been developed. In this work, it is assumed that only bounds on uncertain parameters are available and intervals are used to model uncertainty. Here, we present a new approach to treat uncertainty in supporting conditions. Within the context of Interval Finite Element Method (IFEM), all uncertain parameters are modeled as intervals. However, supporting conditions are considered in idealized types and described by deterministic values without accounting for any form of uncertainty. In the current developed approach, uncertainty in supporting conditions is modeled as bounded range of values, i.e., interval value that capture any possible variation in supporting condition within a given interval. Extreme interval bounds can be obtained by analyzing the considered system under the conditions of the presence and absence of the specific supporting condition. A set of numerical examples is presented to illustrate and verify the accuracy of the proposed approach.

R. L. Muhanna (✉) · S. Shahi
Georgia Institute of Technology, Atlanta, GA 30332, USA
e-mail: rafi.muhanna@gatech.edu

S. Shahi
e-mail: shahi@gatech.edu

© Springer Nature Switzerland AG 2020
M. Ceberio and V. Kreinovich (eds.), *Decision Making under Constraints*,
Studies in Systems, Decision and Control 276,
https://doi.org/10.1007/978-3-030-40814-5_20

1 Introduction

Uncertainties have been studied for improving the predictability of mathematical models. Among others, interval finite element method handles uncertainties in interval form due to the lack of knowledge about the system parameters [2, 4]. Within the context of IFEM, all uncertain parameters are defined as intervals. While uncertainty in load, materials, and geometry in structural system have been intensively studied [2, 3, 6, 10, 11], in spite of their importance, uncertainty in supporting conditions barely have had the needed attention. Modeling boundary conditions is one of the most sensitive steps in the formulation of the mathematical model of a given system. Usually, supporting conditions are idealized through simplifying the mechanical behavior of the actual structural supports. Such idealization, in addition to disregarding the actual supporting conditions and uncertainty involved, may lead to inaccurate evaluation of the actual behavior of the structure [1, 8].

The paper is structured as follows. First, to illustrate the proposed interval approach for handling uncertainty in supporting conditions, we briefly present key features of Element-By-Element formulation of interval finite element. Uncertainty bounds of supporting conditions is then obtained from the extreme conditions and incorporated in the formulation. Examples are finally presented and discussed.

2 Formulation

Usually, in structural systems, supporting conditions are described such as roller, pin, or fixed supports, and in some cases might be modeled as partial restraint. In all these cases, supporting conditions are described by deterministic values. For example, a pin support in plane problems considered to prevent two degrees of freedom (displacements), in other words those degrees of freedom are set equal to zero (Fig. 1a). In real life, such conditions rarely can be met, and any variation in these conditions might happen. Thus, in order to capture uncertainty in a supporting condition, we are modeling the condition as interval value that capture any possible variation within a given interval. In the case of a pin support, for example, one degree of freedom can take the values between zero (lower bound, fully restraint) and a second value (upper bound, fully free) corresponding to the absence of the constraint in direction of the considered degree of freedom (Fig. 1b).

The treatment of uncertain supporting conditions can be achieved by expanding our previous formulation of IFEM [10] through incorporating additional linear or rotational elements whose stiffness is interval, that depends on the specific type of the structure (approach 1). Experts can provide the interval value of the element stiffness. Another way to handle this uncertain situation is to solve the system in the absence of the specific constraint, under the condition that the system remains geometrically stable, and calculate the displacement in the direction of the removed constraint (approach 2). The obtained value of the displacement in the direction of

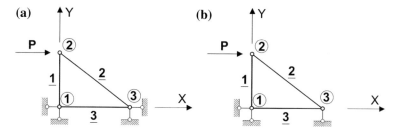

Fig. 1 Truss structure with different boundary conditions. **a** pin-pin at node 1 and 3, **b** pin at node 1 and roller at node 3

the removed constraint can be used as an upper bound of prescribed displacement for which the lower bound is zero.

The same approach can be used to account for uncertainty in the rigidity of the structures joints. When joints connecting structural members are assumed to be rigid, the relative rotation of the adjacent sections of the intersecting members is considered equal to zero. By introducing the rigidity condition as interval, we can consider more realistic behavior of these joints. All interval quantities will be introduced in non-italic boldface font.

Our formulation of interval finite element method is based on the Element-By-Element formulation [6] where each element has its own set of nodes, but the set of elements is disassembled (Fig. 2). A set of additional constraints is imposed to satisfy the conditions of compatibility, these constraints have the form of $CU = V$, where C, U, and V are a constraint matrix, displacement vector, and prescribed displacement values if any, respectively. The constraint matrix describes the connectivity conditions between local element degrees of freedom and corresponding global nodal degrees of freedom in addition to the imposed boundary conditions. For instance, for a joint demonstrated in Fig. 2, the connectivity matrix has the following form,

$$C = \begin{bmatrix} \cdots & \cdots & \cdots & \cdots & & \cdots & & \cdots & & \cdots \cdots \\ \cdots & 1 & 0 & 0 & \cdots & -\cos(\phi) & -\sin(\phi) & 0 & \cdots \\ \cdots & 0 & 1 & 0 & \cdots & \sin(\phi) & -\cos(\phi) & 0 & \cdots \\ \cdots & 0 & 0 & 1 & \cdots & 0 & 0 & 1 & \cdots \\ \cdots & \cdots & \cdots & \cdots & & \cdots & & \cdots & \cdots \cdots \end{bmatrix} \quad (1)$$

The matrix is obtained based on the following conditions

$$u_j - U_k \cos(\phi) - V_k \sin(\phi) = 0$$
$$v_j + U_k \sin(\phi) - V_k \cos(\phi) = 0 \quad (2)$$
$$\theta_j - \Theta_k = 0$$

where ϕ is the angle of rotation between the local and global coordinate systems, u_j, v_j and θ_j are local horizontal, vertical and rotational degrees of freedom of node

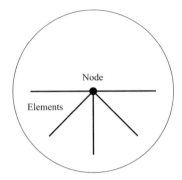

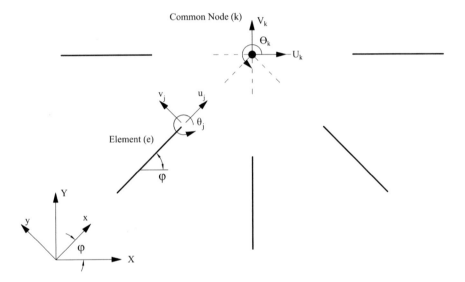

Fig. 2 Element-By-Element. Connection between element nodes and common nodes

j that belongs to element e, respectively, and U_k, V_k and Θ_k are global horizontal, vertical and rotational degrees of freedom of the common node k. Then, the essential boundary conditions are imposed by setting the corresponding components in C equal to 1.

Under these constraints, the amended total potential energy using the Lagrange multiplier approach takes the following form:

$$\Pi^* = \frac{1}{2}U^T K U - U^T P + \lambda^T (CU - V) \tag{3}$$

where Π^*, K, U, P and λ are total potential energy, stiffness matrix, displacement vector, load vector, and Lagrange multipliers vector, respectively. Invoking the stationarity of Π^*, that is $\delta \Pi^* = 0$, we obtain the following interval equilibrium equations, where in the present study, uncertainty is assumed only in the right-hand side,

$$
\begin{bmatrix} \mathbf{K} & C^T \\ C & 0 \end{bmatrix} \begin{bmatrix} \mathbf{U} \\ \lambda \end{bmatrix} = \begin{bmatrix} \mathbf{P} \\ \mathbf{V} \end{bmatrix}. \tag{4}
$$

The advantage of this formulation is that it allows handling any prescribed interval displacement as a part of the right-hand side, i.e., the load vector. Eliminating overestimation due to interval dependency in the load vector has been achieved in our previous work [3].

In approach 1, simply we can add a linear or rotation element with interval property in direction of the considered support accounting for the level of uncertainty. In approach 2, the system is solved after removing the constraint with uncertainty, the obtained displacement value in the direction of the constraint represents the upper bound of the uncertain supporting condition and the lower bound will be equal to zero. This interval range of displacement in direction of the selected degree of freedom is used as prescribed interval displacement in the corresponding entry of V in Eq. 4.

In the case of uncertainty in joint rigidity (partially restrained connections) [5, 7], the system is solved for two extreme deterministic cases: (i) The system with all joints considered rigid, and (ii) the system with hinges introduced at the joints with uncertain rigidity. By deterministically solving (i) and (ii), the rotation of the rigid joint and the rotations at each side of the hinge are calculated, respectively. These values are used to determine the bounds of intervals that model the joints' uncertainties. The obtained intervals are used as prescribed rotations in the right-hand side vector of the equilibrium system of Eq. 4. In the next section, a number of examples are introduced to illustrate the numerical results of the developed formulation.

This formulation for handling uncertainty in supporting conditions and joints' rigidity is an integral part of the previously developed interval finite element that handles uncertainty in materials, geometry, and different loading conditions.

3   Example Problems

The proposed method is implemented in a MATLAB program using the interval toolbox INTLAB to handle interval computations [9]. Three example problems are chosen to illustrate and verify the present approach; a truss and two frames. The first example is an eleven-bar truss as shown in Fig. 3. The numerical results of this example are obtained to illustrate the effect of uncertainty in boundary conditions. The second example is one story single-bay frame with uncertainty in a horizontal boundary condition. This simple example is demonstrating the general applicability of the proposed approach in different types of structures. The third numerical example

Fig. 3 Eleven-bar truss, pin
supported at nodes 1 and 5

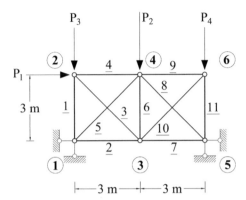

is the same frame structure analyzed in the second example but we illustrate the effect
of the uncertainty in joint rigidity.

These examples do not include any material or load uncertainty, which can be
handled straight forward by the previously developed IFEM. However, the numerical
results illustrate how the proposed extension to IFEM approach provides exact values
of system response under uncertainty in structures' supporting conditions and joints'
rigidity.

3.1 Example 1: Planar Truss with Uncertainty in Boundary Condition

The eleven-bar truss shown in Fig. 3 is subjected to a horizontal concentrated load
$P_1 = 50$ kN at node 2, and three downward concentrated loads of $P_2 = 100$ kN and
$P_3 = P_4 = 50$ kN applied at nodes 4, 2 and 6, respectively. The cross-sectional areas
and the nominal value of Young's modulus of all elements are $A_i = 0.01 \, \text{m}^2$ and
$E_i = 2 \times 10^{11}$ N/m^2, respectively. The dimensions are also illustrated in the figure.

To investigate uncertainty in the horizontal constraint of node 5, the problem is
solved in the absence of that specific constraint (Fig. 4) to obtain the upper bound of
prescribed displacement of that constraint.

Table 1 presents displacement values obtained by solving these two deterministic
cases; namely, pin-pin and pin-roller supporting conditions.

Therefore, the interval prescribed horizontal displacement of the support at node
5 is $[0, 0.18519] \times 10^{-3}$ m. By imposing this condition in the corresponding entry of
vector V as prescribed displacement, the interval solution is obtained. Table 2 presents
the numerical values of displacements due to 100% uncertainty in the horizontal
constraint of node 5.

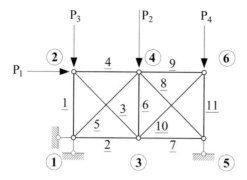

Fig. 4 Eleven-bar truss, roller support at node 5

Table 1 Truss nodal displacements, pin-pin and pin-roller supporting conditions

Node	Pin-Pin		Pin-Roller	
	u (mm)	v (mm)	u (mm)	v (mm)
1	0.00000	0.00000	0.00000	0.00000
2	0.12789	−0.06873	0.23008	−0.07832
3	0.01658	−0.16232	0.10918	−0.20861
4	0.05916	−0.18295	0.15175	−0.24842
5	0.00000	0.00000	0.18519	0.00000
6	0.03226	−0.10190	0.11526	−0.11149

Table 2 Interval displacements due to 100% uncertainty in the support at node 5

Node	u (mm)		v (mm)	
	Lower	Upper	Lower	Upper
1	0.00000	0.00000	0.00000	0.00000
2	0.12789	0.23008	−0.07832	−0.06873
3	0.01658	0.10918	−0.20861	−0.16232
4	0.05916	0.15175	−0.24842	−0.18295
5	0.00000	0.18519	0.00000	0.00000
6	0.03226	0.11526	−0.11149	−0.10190

The results indicate that the bounds of nodal displacement match exactly the displacement results obtained for extreme values given in Table 1. This observation holds true for internal and reaction forces. For instance, the axial force in elements 6 and 7 denoted by N_6 and N_7, respectively, and the horizontal reaction forces at node 1 and 5 denoted by R_1 and R_5, respectively, are presented in Table 3.

Table 3 Selected internal and reaction forces of the truss in Fig. 4 due to 100% uncertainty

Forces (kN)	Deterministic		Interval	
	Pin	Roller	Lower	Upper
N_6	−13.75631	−26.54092	−26.54092	−13.75631
N_7	−11.05606	50.67348	−11.05606	50.67348
R_1	18.12184	−50.00000	−50.00000	18.12184
R_5	−68.12184	0.00000	−68.12184	0.00000

3.2 Example 2: Planar Frame with Uncertainty in Boundary Condition

The next example is a single-story frame subjected to a horizontal concentrated load $P = 5$ kN applied at node 2. The modulus of elasticity, cross sectional area and moment of inertia of all elements are equal to $E = 2 \times 10^{11}$ N/m², $I = 3 \times 10^{-4}$ m⁴ and $A = 8.37 \times 10^{-3}$ m², respectively. In this frame, the effect of uncertainty in the horizontal support at node 4 is studied. To calculate the prescribed displacement interval, the horizontal support at this node is removed by replacing the fixed support with a horizontal slider support (Fig. 5) and the deterministic solution is obtained. Table 4 presents the results of two cases: fixed support at node 4 and horizontal slider support at node 4.

Therefore, the associated interval prescribed horizontal displacement of support at node 4 is $[0, 1.73611] \times 10^{-3}$ m. The obtained interval solution is an exact enclosure of the deterministic solutions of the two extreme cases: Fixed support and slider support at node 4. Numerical results are introduced in Table 5, where u_2 and θ_3 denote the horizontal displacement and rotation of node 2 and 3, respectively. The forces N_2 and M_2 are the axial force and bending moment at the right-end of element 2, respectively, and RM_4 denotes the reaction moment at node 4.

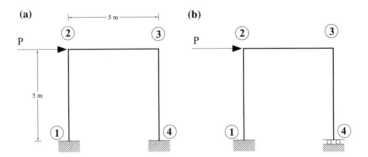

Fig. 5 Planar frame with uncertainty in support condition. **a** Fixed supports at nodes 1 and 4. **b** Fixed at node 1 and slider support at node 4

Table 4 Frame nodal displacements pin-pin and pin-roller supporting conditions

Node	Fixed-Fixed			Fixed-Slider		
	u ($\times 10^{-3}$ m)	v ($\times 10^{-3}$ m)	θ ($\times 10^{-3}$ rad)	u ($\times 10^{-3}$ m)	v ($\times 10^{-3}$ m)	θ ($\times 10^{-3}$ rad)
1	0.00000	0.00000	0.00000	0.00000	0.00000	0.00000
2	0.62922	0.00637	−0.07733	1.49355	0.00637	−0.25020
3	0.62178	−0.00637	−0.07584	1.49355	−0.00637	0.09702
4	0.00000	0.00000	0.00000	1.73611	0.00000	0.00000

Table 5 Effect of 100% uncertainty in horizontal support of frame of Fig. 5a

Forces (kN)	Deterministic		Interval	
	Fixed	Slider	Lower	Upper
$u_2(\times 10^{-3}$ m)	0.62922	1.49355	0.62922	1.49355
$\theta_3(\times 10^{-3}$ rad)	−0.07584	0.09702	-0.07584	0.09702
N_2(kN)	−2.48929	0.00000	−2.48929	0.00000
M_2(kN-m)	−5.31309	−1.16427	−5.31309	−1.16427
RM_4(kN-m)	7.13337	−1.16427	−1.16427	7.13337

Numerical results show the large variability in displacements, internal forces, and reactions due to uncertainty in supporting conditions.

3.3 Example 3: Planar Frame with Uncertainty in Joint Rigidity

This example represents a case study of what is known in structural design as partially restrained connections [5, 7]. The same frame of Fig. 5 is used to illustrate this case. The prescribed interval values at node 2 are obtained by replacing the rigid joint at node 2 by a hinge (Fig. 6b).

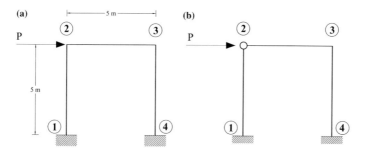

Fig. 6 Planar frame with uncertainty in joint rigidity. **a** Original **b** Hinge at node 2

Table 6 Deterministic rotations of node 2 for the rigid and hinge cases of frames in Fig. 6

Node	Rigid θ ($\times 10^{-3}$ rad)	Hinge θ ($\times 10^{-3}$ rad)
1	0.00000	0.00000
2	−0.07733	(−0.31980, 0.08849)
3	−0.07584	−0.18163
4	0.00000	0.00000

Table 7 Effect of 100% uncertainty in joint 2 rigidity on displacements and forces of the frame in Fig. 6

Forces (kN)	Deterministic		Interval	
	Rigid	Hinge	Lower	Upper
$u_2(\times 10^{-3}$ m)	0.62922	1.06598	0.62922	1.06598
$\theta_3(\times 10^{-3}$ rad)	−0.07584	−0.18163	−0.18163	−0.07584
N_2(kN)	−2.48929	−3.46498	−3.46498	−2.48929
M_2(kN-m)	−5.34878	0.00000	0.00000	5.34878
RM_4(kN-m)	7.13337	10.84202	7.13337	10.84202

Solving the two extreme deterministic cases; the displacements and rotations for the structure with rigid joints and the structure with a hinge at node 2 are obtained. In the case of a hinge at node 2, two different rotation values are obtained for the sections adjacent to the hinge, see Table 6. Thus, the prescribed interval rotations are obtained as $[−0.31980, −0.07733] \times 10^{-3}$ rad and $[−0.0773, 0.08849] \times 10^{-3}$ rad, which are imposed in the corresponding entries of vector V as prescribed displacements.

The selected results are presented in Table 7. The comparison between interval and deterministic solutions reveals that the interval solution is an exact enclosure of the deterministic analysis of extreme cases.

4 Conclusion

This paper presents a new approach in interval finite element method to study the effect of uncertainty in boundary conditions and joint rigidity. This approach treats the uncertainty in supporting conditions as prescribed interval displacements defined in the right-hand side of the system of equations. Exact enclosures of the correct solution are obtained and any form of overestimation is eliminated. The numerical examples illustrate that the proposed approach can successfully capture the effect of uncertainty in supporting conditions and joints rigidity. Numerical results show the large variability in displacements, internal forces, and reactions due to uncertainty in supporting conditions. In addition, this formulation represent an integral part of the IFEM developed previously by the author, and allows to consider uncertainties in materials, geometry, loads, and boundary conditions.

Acknowledgements This work is supported by the National Science Foundation under grant No. 1634483.

References

1. Bursi, O.S., Abbiati, G., Caracoglia, L., Reza, M.S.: Effects of uncertainties in boundary conditions on dynamic characteristics of industrial plant components. In: ASME 2014 Pressure Vessels and Piping Conference. V008T08A028–V008T08A028 (2014)
2. Muhanna, R.L., Mullen, R.L.: Uncertainty in mechanics problems—interval-based approach. J. Eng. Mech. **6**, 557–566 (2001)
3. Mullen, R.L., Muhanna, R.L.: Bounds of structural response for all possible loading combinations. J. Struct. Eng. **1**, 98–106 (1999)
4. Neumaier, A., Pownuk, A.: Linear systems with large uncertainties, with applications to truss structures. Reliab. Comput. **2**, 149–172 (2007)
5. Nielson, B.G., McCormac, J.C.: Structural Analysis: Understanding Behavior. Wiley, Hoboken (2017)
6. Rao, M.R., Mullen, R.L., Muhanna, R.L.: A new interval finite element formulation with the same accuracy in primary and derived variables. Int. J. Reliab. Saf. **5**, 336–357 (2011)
7. Ricles, J., Bjorhovde, R., Iwankiw, N.: Frames with partially restrained connections. In: Workshop Proceedings Structural Stability Research Council, Atlanta (1998–2000)
8. Ritto, T.G., Sampaio, R., Cataldo, E.: Timoshenko beam with uncertainty on the boundary conditions. J. Braz. Soc. Mech. Sci. Eng. **4**, 295–303 (2008)
9. Rump, S.M.: INTLAB - INTerval LABoratory. In: Csendes, Tibor (ed.) Developments in Reliable Computing, pp. 77–104. Kluwer Academic Publishers, Dordrecht (1999)
10. Xiao, N., Muhanna, R.L., Fedele, F., Mullen, R.L.: Interval finite elements for uncertainty in frame structures. In: 11th International Conference on Structural Safety & Reliability, New York, pp. 16–20 (2013)
11. Xiao, N., Muhanna, R.L., Fedele, F., Mullen, R.L.: Uncertainty analysis of static plane problems by intervals. SAE Int. J. Mater. Manuf. **8**, 374–381 (2015)

Which Value $\widetilde{x}$ Best Represents a Sample $x_1, \ldots, x_n$: Utility-Based Approach Under Interval Uncertainty

Andrzej Pownuk and Vladik Kreinovich

Abstract In many practical situations, we have several estimates $x_1, \ldots, x_n$ of the same quantity x. In such situations, it is desirable to combine this information into a single estimate $\widetilde{x}$. Often, the estimates x_i come with interval uncertainty, i.e., instead of the exact values x_i, we only know the intervals $[\underline{x}_i, \overline{x}_i]$ containing these values. In this paper, we formalize the problem of finding the combined estimate $\widetilde{x}$ as the problem of maximizing the corresponding utility, and we provide an efficient (quadratic-time) algorithm for computing the resulting estimate.

1 Which Value $\widetilde{x}$ Best Represents a Sample $x_1, \ldots, x_n$: Case of Exact Estimates

Need to combine several estimates. In many practical situations, we have several estimates $x_1, \ldots, x_n$ of the same quantity x. In such situations, it is often desirable to combine this information into a single estimate $\widetilde{x}$; see, e.g., [6].

Probabilistic case. If we know the probability distribution of the corresponding estimation errors $x_i - x$, then we can use known statistical techniques to find $\widetilde{x}$, e.g., we can use the Maximum Likelihood Method; see, e.g., [8].

Need to go beyond the probabilistic case. In many cases, however, we do not have any information about the corresponding probability distribution [6]. How can we then find $\widetilde{x}$?

Utility-based approach. According to the general decision theory, decisions of a rational person are equivalent to maximizing his/her *utility value u*; see, e.g., [1, 4, 5, 7]. Let us thus find the estimate $\widetilde{x}$ for which the utility $u(\widetilde{x})$ is the largest.

A. Pownuk · V. Kreinovich (✉)
University of Texas at El Paso, El Paso, Texas 79968, USA
e-mail: vladik@utep.edu

A. Pownuk
e-mail: ampownuk@utep.edu

© Springer Nature Switzerland AG 2020
M. Ceberio and V. Kreinovich (eds.), *Decision Making under Constraints*,
Studies in Systems, Decision and Control 276,
https://doi.org/10.1007/978-3-030-40814-5_21

Our objective is to use a single value $\tilde{x}$ instead of all n values x_i. For each i, the disutility $d = -u$ comes from the fact that if the actual estimate is x_i and we use a different value $\tilde{x} \neq x_i$ instead, we are not doing an optimal thing. For example, if the optimal speed at which the car needs the least amount of fuel is x_i, and we instead run it at a speed $\tilde{x} \neq x_i$, we thus waste some fuel.

For each i, the disutility d comes from the fact that the difference $\tilde{x} - x_i$ is different from 0; there is no disutility if we use the actual value, so $d = d(\tilde{x} - x_i)$ for an appropriate function $d(y)$, where $d(0) = 0$ and $d(y) > 0$ for $y \neq 0$.

The estimates are usually reasonably accurate, so the difference $x_i - \tilde{x}$ is small, and we can expand the function $d(y)$ in Taylor series and keep only the first few terms in this expansion:

$$d(y) = d_0 + d_1 \cdot y + d_2 \cdot y^2 + \cdots$$

From $d(0) = 0$ we conclude that $d_0 = 0$. From $d(y) > 0$ for $y \neq 0$ we conclude that $d_1 = 0$ (else we would have $d(y) < 0$ for some small y) and $d_2 > 0$, so

$$d(y) = d_2 \cdot y^2 = d_2 \cdot (\tilde{x} - x_i)^2.$$

The overall disutility $d(\tilde{x})$ of using $\tilde{x}$ instead of each of the values $x_1, \ldots, x_n$ can be computed as the sum of the corresponding disutilities

$$d(\tilde{x}) = \sum_{i=1}^{n} d(\tilde{x} - x_i)^2 = d_2 \cdot \sum_{i=1}^{n} (\tilde{x} - x_i)^2.$$

Maximizing utility $u(\tilde{x}) \stackrel{\text{def}}{=} -d(\tilde{x})$ is equivalent to minimizing disutility.

The resulting combined value. Since $d_2 > 0$, minimizing the disutility function is equivalent to minimizing the re-scaled disutility function

$$D(\tilde{x}) \stackrel{\text{def}}{=} \frac{d(\tilde{x})}{d_2} = \sum_{i=1}^{n} (\tilde{x} - x_i)^2.$$

Differentiating this expression with respect to $\tilde{x}$ and equating the derivative to 0, we get

$$\tilde{x} = \frac{1}{n} \cdot \sum_{i=1}^{n} x_i.$$

This is the well-known sample mean.

2 Case of Interval Uncertainty: Formulation of the Problem

Formulation of the practical problem. In many practical situations, instead of the exact estimates x_i, we only know the intervals $[\underline{x}_i, \overline{x}_i]$ that contain the unknown values x_i. How do we select the value x in this case?

Towards precise formulation of the problem. For different values x_i from the corresponding intervals $[\underline{x}_i, \overline{x}_i]$, we get, in general, different values of utility

$$U(\tilde{x}, x_1, \ldots, x_n) = -D(\tilde{x}, x_1, \ldots, x_n),$$

where $D(\tilde{x}, x_1, \ldots, x_n) = \sum_{i=1}^{n} (\tilde{x} - x_i)^2$. Thus, all we know is that the actual (unknown) value of the utility belongs to the interval $[\underline{U}(\tilde{x}), \overline{U}(\tilde{x})] = [-\overline{D}(\tilde{x}), -\underline{D}(\tilde{x})]$, where

$$\underline{D}(\tilde{x}) = \min D(\tilde{x}, x_1, \ldots, x_n), \quad \overline{D}(\tilde{x}) = \max D(\tilde{x}, x_1, \ldots, x_n),$$

and min and max are taken over all possible combinations of values $x_i \in [\underline{x}_i, \overline{x}_i]$.

In such situations of interval uncertainty, decision making theory recommends using Hurwicz optimism-pessimism criterion [2–4], i.e., maximize the value

$$U(\tilde{x}) \stackrel{\text{def}}{=} \alpha \cdot \overline{U}(\tilde{x}) + (1 - \alpha) \cdot \underline{U}(\tilde{x}),$$

where the parameter $\alpha \in [0, 1]$ describes the decision maker's degree of optimism. For $U = -D$, this is equivalent to minimizing the expression

$$D(\tilde{x}) = -U(\tilde{x}) = \alpha \cdot \underline{D}(\tilde{x}) + (1 - \alpha) \cdot \overline{D}(\tilde{x}).$$

What we do in this paper. In this paper, we describe an efficient algorithm for computing the value $\tilde{x}$ that minimizes the resulting objective function $D(\tilde{x})$.

3 Analysis of the Problem

Let us simplify the expressions for $\underline{D}(\tilde{x})$, $\overline{D}(\tilde{x})$, and $D(\tilde{x})$. Each term $(\tilde{x} - x_i)^2$ in the sum $D(\tilde{x}, x_1, \ldots, x_n)$ depends only on its own variable x_i. Thus, with respect to x_i:

- the sum is the smallest when each of these terms is the smallest, and
- the sum is the largest when each term is the largest.

One can easily see that when x_i is in the $[\underline{x}_i, \overline{x}_i]$, the maximum of a term $(\tilde{x} - x_i)^2$ is always attained at one of the interval's endpoints:

- at $x_i = \underline{x}_i$ when $\widetilde{x} \geq \widetilde{x}_i \overset{\text{def}}{=} \dfrac{\underline{x}_i + \overline{x}_i}{2}$ and
- at $x_i = \overline{x}_i$ when $\widetilde{x} < \widetilde{x}_i$.

Thus,

$$\overline{D}(\widetilde{x}) = \sum_{i:\widetilde{x}<\widetilde{x}_i} (\widetilde{x} - \overline{x}_i)^2 + \sum_{i:\widetilde{x}\geq\widetilde{x}_i} (\widetilde{x} - \underline{x}_i)^2.$$

Similarly, the minimum of the term $(\widetilde{x} - x_i)^2$ is attained:

- for $x_i = \widetilde{x}$ when $\widetilde{x} \in [\underline{x}_i, \overline{x}_i]$ (in this case, the minimum is 0);
- for $x_i = \underline{x}_i$ when $\widetilde{x} < \underline{x}_i$; and
- for $x_i = \overline{x}_i$ when $\widetilde{x} > \overline{x}_i$.

Thus,

$$\underline{D}(\widetilde{x}) = \sum_{i:\widetilde{x}>\overline{x}_i} (\widetilde{x} - \overline{x}_i)^2 + \sum_{i:\widetilde{x}<\underline{x}_i} (\widetilde{x} - \underline{x}_i)^2.$$

So, for $D(\widetilde{x}) = \alpha \cdot \underline{D}(\widetilde{x}) + (1 - \alpha) \cdot \overline{D}(\widetilde{x})$, we get

$$D(\widetilde{x}) = \alpha \cdot \sum_{i:\widetilde{x}>\overline{x}_i} (\widetilde{x} - \overline{x}_i)^2 + \alpha \cdot \sum_{i:\widetilde{x}<\underline{x}_i} (\widetilde{x} - \underline{x}_i)^2 +$$

$$(1 - \alpha) \cdot \sum_{i:\widetilde{x}<\widetilde{x}_i} (\widetilde{x} - \overline{x}_i)^2 + (1 - \alpha) \cdot \sum_{i:\widetilde{x}\geq\widetilde{x}_i} (\widetilde{x} - \underline{x}_i)^2. \qquad (1)$$

Towards an algorithm. The presence or absence of different values in the above expression depends on the relation of $\widetilde{x}$ with respect to the values $\underline{x}_i$, $\overline{x}_i$, and $\widetilde{x}_i$. Thus, if we sort these $3n$ values into a sequence $s_1 \leq s_2 \leq \cdots \leq s_{3n}$, then on each interval $[s_j, s_{j+1}]$, the function $D(\widetilde{x})$ is simply a quadratic function of $\widetilde{x}$.

A quadratic function attains its minimum on an interval either at one of its midpoints, or at a point when the derivative is equal to 0 (if this point is inside the given interval). Differentiating the above expression for $D(\widetilde{x})$, equating the derivative to 0, dividing both sides by 0, and moving terms proportional not containing $\widetilde{x}$ to the right-hand side, we conclude that

$$(\alpha \cdot \#\{i : \widetilde{x} < \underline{x}_i \text{ or } \widetilde{x} > \overline{x}_i\} + 1 - \alpha) \cdot \widetilde{x} =$$

$$\alpha \cdot \sum_{i:\widetilde{x}>\overline{x}_i} \overline{x}_i + \alpha \cdot \sum_{i:\widetilde{x}<\underline{x}_i} \underline{x}_i + (1 - \alpha) \cdot \sum_{i:\widetilde{x}<\widetilde{x}_i} \overline{x}_i + (1 - \alpha) \cdot \sum_{i:\widetilde{x}\geq\widetilde{x}_i} \underline{x}_i.$$

Since s_j is a listing of all thresholds values $\underline{x}_i$, $\overline{x}_i$, and $\widetilde{x}_i$, then for $\widetilde{x} \in (s_j, s_{j+1})$, the inequality $\widetilde{x} < \underline{x}_i$ is equivalent to $s_{j+1} \leq \underline{x}_i$. Similarly, the inequality $\widetilde{x} > \underline{x}_i$ is equivalent to $s_j \geq \overline{x}_i$. In general, for values $\widetilde{x} \in (s_j, s_{j+1})$, the above equation gets the form

$$(\alpha \cdot \#\{i : \widetilde{x} < \underline{x}_i \text{ or } \widetilde{x} > \overline{x}_i\} + 1 - \alpha) \cdot \widetilde{x} =$$

$$\alpha \cdot \sum_{i:s_j \geq \overline{x}_i} \overline{x}_i + \alpha \cdot \sum_{i:s_{j+1} \leq \underline{x}_i} \underline{x}_i + (1-\alpha) \cdot \sum_{i:s_{j+1} \leq \widetilde{x}_i} \overline{x}_i + (1-\alpha) \cdot \sum_{i:s_j \geq \widetilde{x}_i} \underline{x}_i.$$

From this equation, we can easily find the desired expression for the value $\widetilde{x}$ at which the derivative is 0.

Thus, we arrive at the following algorithm.

4 Resulting Algorithm

First, for each interval $[\underline{x}_i, \overline{x}_i]$, we compute its midpoint $\widetilde{x}_i = \dfrac{\underline{x}_i + \overline{x}_i}{2}$. Then, we sort the $3n$ values $\underline{x}_i, \overline{x}_i$, and $\widetilde{x}_i$ into an increasing sequence $s_1 \leq s_2 \leq \cdots \leq s_{3n}$. To cover the whole real line, to these values, we add $s_0 = -\infty$ and $s_{3n+1} = +\infty$.

We compute the value of the objective function (1) on each of the endpoints $s_1, \ldots, s_{3n}$. Then, for each interval (s_i, s_{j+1}), we compute the value

$$\widetilde{x} = \frac{\alpha \cdot \sum_{i:s_j \geq \overline{x}_i} \overline{x}_i + \alpha \cdot \sum_{i:s_{j+1} \leq \underline{x}_i} \underline{x}_i + (1-\alpha) \cdot \sum_{i:s_{j+1} \leq \widetilde{x}_i} \overline{x}_i + (1-\alpha) \cdot \sum_{i:s_j \geq \widetilde{x}_i} \underline{x}_i}{\alpha \cdot \#\{i : \widetilde{x} < \underline{x}_i \text{ or } \widetilde{x} > \overline{x}_i\} + 1 - \alpha}.$$

If the resulting value $\widetilde{x}$ is within the interval (s_i, s_{j+1}), we compute the value of the objective function (1) corresponding to this $\widetilde{x}$.

After that, out of all the values $\widetilde{x}$ for which we have computed the value of the objective function (1), we return the value $\widetilde{x}$ for which objective function $D(\widetilde{x})$ was the smallest.

What is the computational complexity of this algorithm. Sorting $3n = O(n)$ values $\underline{x}_i, \overline{x}_i$, and $\widetilde{x}_i$ takes time $O(n \cdot \ln(n))$.

Computing each value $D(\widetilde{x})$ of the objective function requires $O(n)$ computational steps. We compute $D(\widetilde{x})$ for $3n$ endpoints and for $\leq 3n + 1$ values at which the derivative is 0 at each of the intervals (s_j, s_{j+1}) – for the total of $O(n)$ values.

Thus, overall, we need $O(n \cdot \ln(n)) + O(n) \cdot O(n) = O(n^2)$ computation steps. Hence, our algorithm runs in quadratic time.

Acknowledgements This work was supported in part by the National Science Foundation grants HRD-0734825 and HRD-1242122 (Cyber-ShARE Center of Excellence) and DUE-0926721, and by an award "UTEP and Prudential Actuarial Science Academy and Pipeline Initiative" from Prudential Foundation.

References

1. Fishburn, P.C.: Utility Theory for Decision Making. Wiley, New York (1969)
2. Hurwicz, L.: Optimality criteria for decision making under ignorance. Cowles Comm. Discuss. Pap., Stat. 370 (1951)
3. Kreinovich, V.: Decision making under interval uncertainty (and beyond). In: Guo, P., Pedrycz, W. (eds.) Human-Centric Decision-Making Models for Social Sciences, pp. 163–193. Springer, Berlin (2014)
4. Luce, R.D., Raiffa, R.: Games and Decisions: Introduction and Critical Survey. Dover, New York (1989)
5. Nguyen, H.T., Kosheleva, O., Kreinovich, V.: Decision making beyond Arrow's 'impossibility theorem', with the analysis of effects of collusion and mutual attraction. Int. J. Intell. Syst. **24**(1), 27–47 (2009)
6. Rabinovich, S.G.: Measurement Errors and Uncertainty. Theory and Practice. Springer, Berlin (2005)
7. Raiffa, H.: Decision Analysis. Addison-Wesley, Reading, Massachusetts (1970)
8. Sheskin, D.J.: Handbook of Parametric and Nonparametric Statistical Procedures. Chapman and Hall/CRC, Boca Raton, Florida (2011)

Why Unexpectedly Positive Experiences Make Decision Makers More Optimistic: An Explanation

Andrzej Pownuk and Vladik Kreinovich

Abstract Experiments show that unexpectedly positive experiences make decision makers more optimistic. However, there seems to be no convincing explanation for this experimental fact. In this paper, we show that this experimental phenomenon can be naturally explained within the traditional utility-based decision theory.

1 Formulation of the Problem

Experimental phenomenon. Experiments show that unexpectedly positive experiences make decision makers more optimistic. This was first observed on an experiment with rats [10]: rats like being tickled, and tickled rats became more optimistic. Several later papers showed that the same phenomenon holds for other decision making situations as well; see, e.g., [2, 7].

Similarly, decision makers who had an unexpectedly negative experiences became more pessimistic; see, e.g., [8].

Why: a problem. There seems to be no convincing explanation for this experimental fact.

What we do in this paper. In this paper, we show that this experimental phenomenon can be naturally explained within the traditional utility-based decision theory.

A. Pownuk · V. Kreinovich (✉)
University of Texas at El Paso, El Paso, TX 79968, USA
e-mail: vladik@utep.edu

A. Pownuk
e-mail: ampownuk@utep.edu

© Springer Nature Switzerland AG 2020
M. Ceberio and V. Kreinovich (eds.), *Decision Making under Constraints*,
Studies in Systems, Decision and Control 276,
https://doi.org/10.1007/978-3-030-40814-5_22

2   Formulating the Problem in Precise Terms

In precise terms, what does it mean to becomes more optimistic or less optimistic? The traditional utility-based decision theory describes the behavior of a rational decision maker in situations in which we know the probabilities of all possible consequences of each action; see, e.g., [1, 5, 6, 9]. This theory shows that under this rationality assumption, preferences of a decision maker can be described by a special function $U(x)$ called *utility function*, so that a rational decision maker selects an alternative a that maximizes the expected value $u(a)$ of the utility.

In this case, there is no such thing as optimism or pessimism: we just select the alternative which we know is the best for us.

The original theory describes the behavior of decision makers in situations in which we know the probability of each possible consequence of each action. In practice, we often have only *partial* information about these probabilities—and sometimes, no information at all. In such situations, there are several possible probability distributions consistent with our knowledge. For different distributions, we have, in general, different values of the expected utility. As a result, for each alternative, instead of the exact value of the expected utility, we have an *interval* $[\underline{u}(a), \overline{u}(a)]$ of possible values of $u(a)$. How can we make a decision based on such intervals?

In this case, natural rationality ideas lead to the conclusion that a decision should select an alternative a for which, for some real number $\alpha \in [0, 1]$, the combination $u(a) = \alpha \cdot \overline{u}(a) + (1 - \alpha) \cdot \underline{u}(a)$ is the largest possible; see, e.g., [4]. This idea was first proposed by the Nobelist Leo Hurwicz in [3].

The selection of α, however, depends on the person. The value $\alpha = 1$ means that the decision maker only takes into account the best possible consequences, and completely ignores possible less favorable situations. In other words, the values $\alpha = 1$ corresponds to complete optimism.

Similarly, the value $\alpha = 0$ means that the decision maker only takes into account the worst possible consequences, and completely ignores possible more favorable situations. In other words, the value $\alpha = 0$ corresponds to complete pessimism.

Intermediate values α mean that we take into account both positive and negative possibilities. The larger α, the close this decision maker to complete optimism. The smaller α, the closer the decision maker to complete pessimism. Because of this, the parameter α—known as the *optimism-pessimism index*—is a numerical measure of the decision maker's optimism.

In these terms:

- becoming more optimistic means that the value α increases, and
- becoming less optimistic means that the value α increases.

Thus, the above experimental fact takes the following precise meaning:

- if a decision maker has unexpectedly positive experiences, then this decision maker's α increases;
- if a decision maker has unexpectedly negative experiences, then this decision maker's α decreases.

This is the phenomenon that we need to explain.

3 Towards the Desired Explanation

Optimism-pessimism parameter α can be naturally interpreted as the subjective probability of positive outcome. The value α means that the decision maker selects an alternative a for which the value $\alpha \cdot \overline{u}(a) + (1 - \alpha) \cdot \underline{u}(a)$ is the largest possible.

Here, the value $\overline{u}(a)$ corresponds to the positive outcome, and the value $\underline{u}(a)$ corresponds to negative outcome.

For simplicity, let us consider the situation when we have only two possible outcomes:

- the positive outcome, with utility $\overline{u}(a)$, and
- the negative outcome, with utility $\underline{u}(a)$.

A traditional approach to decision making, as we have mentioned, assumes that we know the probabilities of different outcomes. In this case of uncertainty, we do not know the actual (objective) probabilities, but we can always come up with estimated (subjective) ones.

Let us denote the subjective probability of the positive outcome by p_+. Then, the subjective probability of the negative outcome is equal to $1 - p_+$. The expected utility is equal to $p_+ \cdot \overline{u}(a) + (1 - p_+) \cdot \underline{u}(a)$.

This is exactly what we optimize when we use Hurwicz's approach, with $\alpha = p_+$. Thus, the value α can be interpreted as the subjective probability of the positive outcome.

A new reformulation of our problem. In these terms, the above experimental phenomenon takes the following form:

- unexpectedly positive experiences increase the subjective probability of a positive outcome, while
- unexpectedly negative experiences decrease the subjective probability of a positive outcome.

To explain this phenomenon, let us recall where subjective probabilities come from.

Where subjective probabilities come from? A natural way to estimate the probability of an event is to consider all situations in which this event could potentially happen, and then take the frequency of this event—i.e., the ratio n/N of the number of times n when it happens to the overall number N of cases—as the desired estimate for the subjective probability. For example, if we flip a coin 10 times and it fell heads 6 times out of 10, we estimate the probability of the coin falling heads as 6/10.

Let us show that this leads to the desired explanation.

Resulting explanation. Suppose that a decision maker had n positive experiences in the past N situations. Then, the decision maker's subjective probability of a positive outcome is $p_+ = n/N$.

Unexpectedly positive experiences means that we have a series of new experiments, in which the fraction of positive outcomes was higher than the expected frequency p_+. In other words, unexpectedly positive experiences means that $n'/N' > p$, where N' is the overall number of new experiences, and n' is the number of those new experiences in which the outcome turned out to be positive.

How will these new experiences change the decision maker's subjective probability? Now, the decision maker has encountered overall $N + N'$ situations, of which $n + n'$ were positive. Thus, the new subjective probability p'_+ is equal to the new ratio $p'_+ = \dfrac{n + n'}{N + N'}$. Here, by definition of p_+, we have

$$n = p_+ \cdot N$$

and, due to unexpected positiveness of new experiences, we have $n' > p_+ \cdot N'$. By adding this inequality and the previous equality, we conclude that $n + n' > p_+ \cdot (N + N')$, i.e., that

$$p'_+ = \frac{n + n'}{N + N'} > p_+.$$

In other words, unexpectedly positive experiences increase the subjective probability of a positive outcome.

As we have mentioned, the subjective probability of the positive outcome is exactly the optimism-pessimism coefficient α. Thus:

- the original subjective probability p_+ is equal to the original optimism-pessimism coefficient α, and
- the new subjective probability p'_+ is equal to the new optimism-pessimism coefficient α'.

So, the inequality $p'_+ > p_+$ means that $\alpha' > \alpha$, i.e., that unexpectedly positive experiences make the decision maker more optimistic. This is exactly what we wanted to explain.

Similarly, if we had unexpectedly negative experiences, i.e., if we had $n' < p_+ \cdot N'$, then we similarly get $n + n' < p_+ \cdot (N + N')$ and thus,

$$p'_+ = \frac{n + n'}{N + N'} < p_+$$

and $\alpha' < \alpha$. So, we conclude that unexpectedly negative experiences make the decision maker less optimistic. This is also exactly what we observe. So, we have the desired explanation.

Acknowledgements This work was supported in part by the National Science Foundation grants HRD-0734825 and HRD-1242122 (Cyber-ShARE Center of Excellence) and DUE-0926721, and by an award "UTEP and Prudential Actuarial Science Academy and Pipeline Initiative" from Prudential Foundation.

References

1. Fishburn, P.C.: Utility Theory for Decision Making. Wiley, New York (1969)
2. Hales, C.A., Stuart, S.A., Anderson, M.H., Robinson, E.S.J.: Modelling cognitive affective biases in major depressive disorder using rodents. Br. J. Pharmacol. **171**(20), 4524–4538 (2014)
3. Hurwicz, L.: Optimality criteria for decision making under ignorance. Cowles Commission Discussion Paper, Statistics, No. 370 (1951)
4. Kreinovich, V.: Decision making under interval uncertainty (and beyond). In: Guo, P., Pedrycz, W. (eds.) Human-Centric Decision-Making Models for Social Sciences, pp. 163–193. Springer, Berlin (2014)
5. Luce, R.D., Raiffa, R.: Games and Decisions: Introduction and Critical Survey. Dover, New York (1989)
6. Nguyen, H.T., Kosheleva, O., Kreinovich, V.: Decision making beyond Arrow's 'impossibility theorem', with the analysis of effects of collusion and mutual attraction. Int. J. Intell. Syst. **24**(1), 27–47 (2009)
7. Panksepp, J., Wright, J.S., Döbrössy, M.D., Schlaepfer, ThE, Coenen, V.A.: Affective neuroscience strategies for understanding and treating depression: from preclinical models to three novel therapeutics. Clin. Psychol. Sci. **2**(4), 472–494 (2014)
8. Papciak, J., Popik, P., Fuchs, E., Rygula, R.: Chronic psychosocial stress makes rats more 'pessimistic' in the ambiguous-cue interpretation paradigm. Behav. Brain Res. **256**, 305–310 (2013)
9. Raiffa, H.: Decision Analysis. Addison-Wesley, Reading, MA (1970)
10. Rygula, R., Pluta, H., Popik, P.: Laughing rats are optimistic. PLoS ONE **7**(2), e51959 (2012)

Back to Classics: Controlling Smart Thermostats with Natural Language… with Personalization

Julia Taylor Rayz, Saltanat Tazhibayeva and Panagiota Karava

Abstract Fuzzy Sets and Fuzzy Logic are classical tools for controlling devices, including thermostats. Fuzzy sets have been also used to describe Natural Language hedges that appear in every day speech. We use both of these classic applications for personalized thermal preferences of occupants with a goal of saving energy. As such, we combine previous knowledge on preferred comfort temperature range of groups of individuals with fuzzy hedges for temperature control setting, create fuzzy sets for energy consumption and saving, and use the intersection of the created sets with the preferred temperatures to optimize natural language interpretations of occupants' commands on temperature settings of smart thermostats.

Keywords Natural language · Computing with words · Fuzzy hedges · Smart thermostat

1 Introduction

If one is to type a term *thermostat* in a web brother, the result is likely to be a collection of smart devices. A smart thermostat typically integrates a user-friendly interface that allows graphical or verbal communication with the device. A topic of this paper is verbal communication with smart thermostats. In particular, we are interested not only in the verbal explicit commands that, if everything works, a thermostat is able to process, but also verbal commands that can be imprecise and should be interpreted based on the situation and an environment.

An easy example an interaction that requires interpretation is a command "increase the temperature by a few degrees." Of course, the command is given to a system that

J. T. Rayz (✉) · S. Tazhibayeva · P. Karava
Purdue University, West Lafayette, IN 47907, USA
e-mail: jtaylor1@purdue.edu

S. Tazhibayeva
e-mail: stazhiba@purdue.edu

P. Karava
e-mail: pkarava@purdue.edu

© Springer Nature Switzerland AG 2020
M. Ceberio and V. Kreinovich (eds.), *Decision Making under Constraints*,
Studies in Systems, Decision and Control 276,
https://doi.org/10.1007/978-3-030-40814-5_23

is capable of recognizing it (Amazon Alexa, Google home, etc.), processing the words and translating them to what the thermostat should actually do. This requires knowledge the current temperature, interpreting what *a few* means in this domain, and actually setting the setpoint to the resulting temperature. Arguably, an interpretation of *a few* should vary based on the domain, even in the narrow case of temperature control. For instance, one may wish the resulting change in temperature in both Fahrenheit or Celsius be similar, causing the *a few* in Celsius be approximately half of *a few* in Fahrenheit. On the other hand, it is possible that one would wish to stabilize the definition of *a few*, understanding that it would result in a rather different interpretation of the command.

The interpretation of hedges [1, 2]–such as in "*slightly* increase the temperature"— or quantifiers [3, 4] has been a subject of numerous papers in fuzzy sets [5]. Zadeh [1] defines two types of hedges:

Type I: Hedges in this category can be represented as operators acting on a fuzzy set. Typical hedges in this category are: *very, more or less, much, slightly, highly*.

Type II: Hedges in this category require a description of how they act on the components of the operand. Typical hedges in this category are: *essentially, technically, actually, strictly, in a sense, practically, virtually, regular*, etc.

It is tempting to restrict the use of interaction in the temperature domain to the Type I hedges. However, it may be more practical and more natural to include the Type II hedges as well. For example, consider the following interaction:

- X, can you *slightly* increase the temperature in the dining room?
- I am *practically* at my recommended maximum temperature, are you sure you wish me to go any higher?
- X, please increase it *very slightly*?
- Sure, but you will notice *virtually* no difference.

There are two points that are worth noting here. The first one is that the communication between a user and a computational entity that can operate a thermostat is natural. The second one is that this computational entity is capable of providing feedback and suggestions based on the current setpoint and some predefined or learned behavior. The paper discusses both of these points in some detail.

2   Talking to a Smart Device…

The advances in interaction between humans and computational devices are making huge steps forward every day. However, most devices still require essentially precise instructions for their operations. The odds of a thermostat reacting correctly to "set the temperature to 72 degrees" is (slightly?) higher than that of a "increase the temperature by 3 degrees" and higher than that of an "increase the temperature by a few [several?] degrees." While the satisfaction of the third version of the command depends on an interpretation of what *a few* [6] or *several* mean, in many cases the actual temperature increase is achieved, even though it is not clear whether the words are interpreted as crisp or fuzzy.

In human to human communication, however, it is not necessary to explicitly state the request to "increase the temperature" to achieve the same results. It is just as possible to state "I am cold" for the desired effect to happen. Arguable, increasing the temperature is not the only solution to the problem when two people are talking, especially if many instruments of correction are available. For instance, one may suggest dressing warmer, closing windows, or any other suitable alternative. We will assume here that the most salient interpretation depends on what participants can achieve in the most sensible way. For example, if the window is open, it is cold outside, and it is possible to close the window, it should be closed to solve the "I am cold" problem. However, if one is not talking to a human—who can do anything—but to a robotic or computational system with limited capabilities, the number of choices that are available to solve the problem is greatly reduced. In fact, there are two options that are available to a system that can only talk or control temperature: (1) to suggest as to how to solve the problem, such as "have you considered putting warmer clothes on," or (2) to increase the temperature.

While recognizing the intent of the utterance is outside of the scope of computing hedges or quantifiers, and is somewhat outside of the capabilities of most conversational agents, it is worth mentioning that the request should be interpreted as "increase the temperature to my comfort level." It is this meaning of the *comfort level*, which is very much imprecise, but can be approximated, that takes us to the next level of computing with words [7–10].

We make another assumption, realizing that it may be an unwelcome one, that in order of the personalization to happen, the data about an individual have to be collected. In particular, we assume that we have information about the typical thermostat settings for the individual requesting the temperature to be set at his comfort level. Moreover, we assume that the schedule and the setpoints in the smart thermostat only approximate the preferred temperature for any give mode (*home, away, sleep*, etc.) and thus can be modeled as fuzzy sets.

The modes themselves make the calculation of the temperature slightly more complex, but they do not, per se, complicate the interaction between a human and a smart device. The only change that is required in the execution of the request above is the introduction of the fuzzy rules similar to the following:

- If person is cold (and is at home talking to the device), and mode is currently set to *away*, change it to *home.*
- If mode is set to *sleep*, and person is cold (and is talking, thus awake), change the mode to *home.*
- If mode is set to *home,* and person is cold, increase the temperature.

It is the last rule that we will address in the next section as is it arguably the trickiest one: supposedly, *home* mode is set to the most comfortable level by a competent user. If a user is not as competent as one would wish, then the solution to the rule is somewhat easier: calculate his preferred temperature when he is home and set the temperature to it. Let us assume for now that we work with a competent user and the request is legitimate: the temperature is set to the preferred setting, but the person is still cold and needs the temperature to be higher.

Let us also consider another scenario for a competent user: the user is cold, but instead of simply stating this fact, (s)he asks the conversational agent to "increase the temperature by several degrees." Linguistically, the statements are very different: "I am cold" (read as "increase the temperature by an unspecified number of degrees") and "increase the temperature by several degrees" (a number is specified, but still imprecise). In practice, for an energy aware thermostat (and its connected conversational agent), the problem is reduced to a calculation of the lowest satisfiable temperature for this increase.

A very similar problem would be encountered for the cooling system, with the decrease of temperature being requested, and the person likely to be too hot, or heat-cool dual mode that would present more rules but very similar solutions.

3   Energy-Aware Temperature Adjustment Request

Lee et al. [11] examined human thermal preferences in the office buildings and developed smart environment control system based on these preferences. They then tested thermal preference profiles of new occupants based on the sub-models developed for each of the clusters of people who have similar preference characteristics. Figure 1 shows a fragment of data, representing probability distribution calculated with sub-models for different clusters.

Awalgaonkar et al. [12] showed that it is possible to learn individual thermal preferences of people by posing 5–10 questions about their comfort level overtime. Combining these two results, we conclude that it is possible to learn both individual and group preferences of people and use them in adjusting temperature settings.

Fig. 1 The probability distributions calculated with sub-models. A fragment of original figure [11]

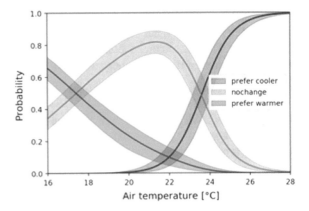

We propose to treat an individual or a group thermal preference as fuzzy sets modeled by Gaussian distribution. We further propose to use known apparatus of linguistic hedges/quantifiers to apply them to temperature setting scenario. We will proceed with two examples, as outlined in the previous section.

3.1 Example 1: "Increase the Temperature by Several Degrees"

We will model the word *several* as a trapezoidal membership function $(t, t + a, b, b + a)$, where t is the current temperature:

$$\mu_{several}(x) = \begin{cases} 0, x < t \text{ or } x > b + a \\ \frac{x-t}{a}, t \le x \le t + a \\ 1, a + t \le x \le b \\ \frac{b+a-x}{a+b}, b \le x \le b + a \end{cases}$$

We also assume that a comfortable can be modeled by a Gaussian membership function:

$$\mu_{comfortable}(x) = e^{-\frac{(c-x)^2}{2\sigma^2}}$$

We propose that instead of using the fuzzy set *several* independently of the *comfort* level when adjusting the temperature, both sets are considered by finding the maximum temperature in the intersection of the two membership sets:

$$\mu_{comfortable\ while\ increased\ by\ several\ degrees}(x) = \min\{\mu_{comfortable}(x), \mu_{several}(x)\};$$
$$t_{new} = \max(\min\{\mu_{comfortable}(x), \mu_{several}(x)\})$$

The graphical illustration is demonstrated in Fig. 2: suppose the current temperature set at 65°, and a user requests to increase the temperature by several degrees (*several* set is shown in orange). If the system only considered the existing definition of several, it would stop far short from the comfort level of a user, shown in blue. On the other hand, combining both sets results in the new temperature of close to 72°. Similarly, if the current temperature is set to 70°, it is possible that the system would increase the temperature to more than 72 (which corresponds to the highest membership of comfort).

Fig. 2 User's comfort set, shown in blue; fuzzy set corresponding to a several-degree increase when the current temperature is 65° (shown in orange); fuzzy set corresponding to a several-degree increase when the current temperature is set to 70 (shown in green)

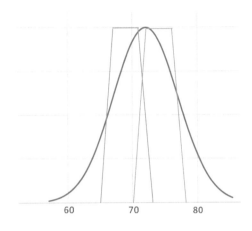

3.2   Example 2: "I Am Cold"

In this example, we again assume that we are familiar with the user's comfort level, the temperature is set at the comfort level, yet it is not sufficient, as illustrated in Fig. 3. For convenience of demonstration we can assume that an increase of temperature will be governed by a membership function similar to *several*, although it is likely that *a few* may be tried as well. We use the same equations as in the example above, resulting in the new temperature of approximately 74°. This is lower than temperature of *several* with membership value of 1 (shown in red), thus if the user is satisfied with the temperature, we save some energy. If the user is not satisfied, we continue the increase the temperature using the same technique (purple set in Fig. 3). Notice that this temperature is not optimal according to the user's normal comfort level, but it does continue to increase until (s)he is satisfied.

Fig. 3 User's comfort set, shown in blue; fuzzy set corresponding to a several-degree increase when the current temperature is 72° (shown in red); if the user requests an increase again, the new temperature will correspond to the purple dotted line

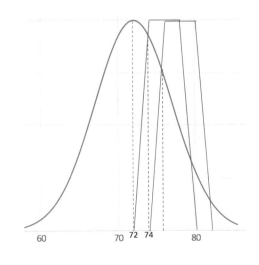

4 Conclusion

This short paper outlines how fuzzy techniques that have been continuously applied in Computing with Words can used for informal communication with smart devices (more established versions of which relied on fuzzy logic quite a bit as well). We show that it is possible to customize communication with smart thermostats by understanding user comfort levels and adjust requests for temperature changes based on both standardly available fuzzy functions for words and individual preferences in temperature. We propose to rely on such customization in order to save energy as well as achieving a high comfort level of individuals.

Acknowledgements This work has been partially supported by National Science Foundation under grant number 1737591.

References

1. Zadeh, L.A.: A fuzzy set theoretic interpretation of linguistics hedges. J. Cybern. **2**(3), 4–34 (1972)
2. Lakoff, G.: Hedge: a study in meaning criteria and the logic of fizzy concepts. J. Philos. Log. **2**(4), 458–508 (1973)
3. Glöckner, I.: Fuzzy Quantifiers: A Computational Theory. Springer (2006)
4. Delgado, M., Ruiz, M.D., Sanchez, D., Vila, M.A.: Fuzzy quantification: a state of the art. Fuzzy Sets Syst. **242**, 1–30 (2014)
5. Zadeh, L.: Fuzzy sets. Inf. Control **8**, 338–353 (1965)
6. Zadeh, L.A.: A computational approach to fuzzy quantifiers in natural languages. Comput. Math Appl. **9**(1), 149–184 (1983)
7. Zadeh, L.A.: From computing with numbers to computing with words. From manipulation of measurements to manipulation of perceptions. IEEE Trans. Circuits Syst. I Fundam. Theory Appl. **46**(1) (1999)
8. Zadeh, L.A.: Computing with Words: Principal Concepts and Ideas. Springer (2012)
9. Mendel, J.M.: Computing with words, when words can mean different things to different people. In: ICSC Congress on Computational Intelligence: Methods and Applications (1999)
10. Raskin, V., Taylor, J.: Computing with nouns and verbs. In: IEEE International Conference on Fuzzy Systems (2012)
11. Lee, S., Karava, P., Tzempelikos, A., Bilionis, I.: Inference of thermal preference profiles for personalized thermal environments with actual building occupants. Build. Environ. (2018)
12. Awalgaonkar, N., Bilionis, I., Liu, X., Karava, P., Tzempelikos, A.: Learning personalized thermal preferences via Bayesian active learning with unimodality constraints. http://arxiv.org/abs/1903.09094 (2019)

The Role of Affine Arithmetic in Robust Optimal Power Flow Analysis

Alfredo Vaccaro

Abstract Optimal Power Flow (OPF) analysis represents the mathematical foundation of many power engineering applications. For the most common formalization of the OPF problem, all input data are specified using deterministic variables, and the corresponding solutions are deemed representative of the limited set of system conditions. Hence, reliable algorithms aimed at representing the effect of data uncertainties in OPF analyses are required in order to allow analysts to estimate both the data and solution tolerance, providing, therefore, insight into the level of confidence of OPF solutions. To address this issue, this Chapter outline the role of novel solution methodologies based on the use of Affine Arithmetic.

Optimal power system operation requires intensive numerical analysis to study and improve system security and reliability. In this context, power system operators need to understand and reduce the impact of system uncertainties. To address this issue, Optimal Power Flow (OPF) analysis is one of the most important tool, since it represents the mathematical foundation of many power engineering applications such as state estimation, network optimization, unit commitment, voltage control, generation dispatch, and market studies.

For the most common formalization of the OPF problem, all input data are specified using deterministic variables resulting either from a snapshot of the system or defined by the analyst based on several assumptions about the system under study (e.g. expected/desired generation/load profiles). This approach provides OPF solutions for a single system state that is deemed representative of the limited set of system conditions corresponding to the data assumptions. Thus, when the input conditions are uncertain, numerous scenarios need to be analyzed. These uncertainties are due to several internal and external sources in power systems. The most relevant

A. Vaccaro (✉)
Department of Engineering, University of Sannio, Piazza Roma 21,
82100 Benevento, Italy
e-mail: vaccaro@unisannio.it

© Springer Nature Switzerland AG 2020
M. Ceberio and V. Kreinovich (eds.), *Decision Making under Constraints*,
Studies in Systems, Decision and Control 276,
https://doi.org/10.1007/978-3-030-40814-5_24

uncertainties are related to the complex dynamics of the active and reactive power supply and demand, which may vary due to, for example:

- the variable nature of generation patterns due to competition [1];
- the increasing number of smaller geographically dispersed generators that could sensibly affect power transactions [1];
- the difficulties arising in predicting and modeling market operators behavior, governed mainly by unpredictable economic dynamics, which introduce considerable uncertainty in short-term power system operation; and
- the high penetration of generation units powered by non-dispatchable renewable energy sources that induce considerable uncertainty in power systems operation [2].

Since uncertainties could affect the OPF solution to a considerable extent, reliable solution paradigms, incorporating the effect of data uncertainties, are required. Such algorithms could allow analysts to estimate both the data tolerance (i.e. uncertainties characterization) and the solution tolerance (i.e. uncertainty propagation assessment), providing, therefore, insight into the level of confidence of OPF solutions. Furthermore, these methodologies could effectively support sensitivity analysis of large parameters variations to estimate the rate of change in the solution with respect to changes in input data.

To address the aforementioned problem, this Chapter analyzes novel solution methodologies based on the use of Affine Arithmetic, which is an enhanced model for self-validated numerical analysis in which the quantities of interest are represented as affine combinations of certain primitive variables representing the sources of uncertainty in the data or approximations made during computations. Compared to existing solution paradigms, this formulation presents greater flexibility, as it allows to find partial solutions and inclusion of multiple equality and inequality constraints, and reduce the approximation errors to obtain better OPF solution enclosures.

Detailed numerical results obtained on a real case study are presented and discussed, demonstrating the effectiveness of the proposed methodologies, especially in comparison to more traditional techniques.

1 Motivations

Conventional methodologies available in the literature propose the use of sampling, analytical and approximate methods for OPF analysis [3, 4], accounting for the variability and stochastic nature of the input data used. A critical review of the most relevant papers proposing these solution methodologies is presented in the following subsections.

1.1 Sampling Methods

Uncertainty propagation studies based on sampling based methods, such as Monte Carlo, require several model runs that sample various combinations of input values. In particular, the most popular Monte Carlo based algorithm adopted to solve OPF problems is simple random sampling, in which a large number of samples are randomly generated from the probability distribution functions of the input uncertain variables. Although this technique can provide highly accurate results, it has the drawback of requiring high computation resources needed for the large number of repeated OPF solutions [5]. This hinders the application of this solution algorithm, especially for large scale power system analysis, where the number of simulations may be rather large and the needed computational resources could be prohibitively expensive [6].

The need to reduce the computational costs of Monte Carlo simulations, has stimulated the research for improved sampling techniques aimed at reducing the number of model runs, at the cost of accepting some level of risk. For example, in [7], an efficient Monte Carlo method integrating Latin hypercube sampling and Cholesky decomposition is proposed to solve PF problems. In [8], the uncertain PF problem with statistically correlated input random variables is solved by a hybrid solution algorithm based on deterministic annealing expectation maximization algorithm and Markov chain Monte Carlo. An extended Latin hypercube sampling algorithm aimed at solving PF problems in the presence of correlated wind generators is proposed in [9]. In [10], the uncertain OPF problem is formulated as a chance-constrained programming model, and the stochastic features of its solutions are obtained by combining Monte Carlo based simulations with deterministic optimisation models.

Although the application of the aforementioned techniques allow to lower the computational burden of sampling-based approaches, these reduce the accuracy of the estimation of uncertainty regions of PF and OPF solutions. Therefore, the dichotomy between accuracy and computational efficiency is still an open problem that requires further investigation.

1.2 Analytical Methods

Analytical methods are computationally more effective, but they require some mathematical assumptions in order to simplify the problem and obtain an effective characterization of the output random variables [11]. These assumptions are typically based on model multi-linearization [12], convolution techniques, and fast Fourier transforms [13]. For example, the cumulant method has been applied to solve the probabilistic OPF problem in [11]; the performance of this method is enhanced by combining it with the Gram-Charlier expansion in [14], and by integrating the Von Mises functions in [15], to handle discrete distributions.

Analytical techniques present various shortcomings, as discussed in [16–18], such as the need to assume statistical independence of the input data, and the problems associated with accurately identifying probability distributions for some input data.

This is a problem for PF and OPF analysis, since it is not always feasible to translate imprecise knowledge into probability distributions, as in the case of power generated by wind or photovoltaic generators, due to the inherently qualitative knowledge of the phenomena and the lack of sufficient data to estimate the required probability density distributions. To address this issue, the assumptions of normality and statistical independence of the input variables are often made, but experimental results show that these assumptions are often not supported by empirical evidence. These drawbacks may limit the usefulness of analytical methods in practical applications, especially for the study of large-scale power networks.

1.3 Approximate Methods

In order to overcome some of the aforementioned limitations of sampling and analytical methods, the use of approximate methods, such as the first-order second-moment method and point estimate methods, have been proposed in the literature [19]. Rather than computing the exact OPF solution, these methods aim at approximating the statistical proprieties of the output random variables by means of a probability distribution fitting algorithm. In particular, the application of the first-order second-moment method allows to compute the first two moments of the OPF solution by propagating the moments of the input variables by the Taylor series expansion of the model equations [20].

Approximate solution methods present several shortcomings. In particular, two-point estimate methods are not suitable to solve large scale problems, since they typically do not provide acceptable results in the presence of a large number of input random variables. Moreover, the identification of the most effective scheme that should be adopted to select the number of estimated points is still an open problem that requires further investigations [21]; this is a critical issue, since a limited number of estimated points does not allow for an accurate and reliable exploration of the solution space, especially for input uncertainties characterized by relatively large standard deviations, such as in the case of lognormal or exponential distributions [21]. On the other hand, an increased number of estimated points reduces the computational benefits deriving by the application of point estimated methods, which could degenerate into a standard Monte Carlo solution approach.

1.4 Non-probabilistic Methods

Recent research has enriched the spectrum of available techniques to deal with uncertainty in OPF by proposing self-validated computing for uncertainty representation in OPF analysis. The main advantage of self-validated computation is that the algorithm itself keeps track of the accuracy of the computed quantities, as part of the process of computing them, without requiring information about the type of uncertainty in the parameters [22]. The simplest and most popular of these models is

Interval Mathematics (IM), which allows for numerical computation where each quantity is represented by an interval of floating point numbers without a probability structure [23]. Such intervals are added, subtracted, and/or multiplied in such a way that each computed interval is guaranteed to contain the unknown value of the quantity it represents.

The application of "standard" IM, referred here as interval arithmetic (IA), to PF analysis has been investigated by various authors [17, 18, 24, 25]. However, the adoption of this solution technique present many drawbacks derived mainly by the so called "dependency problem" and "wrapping effect" [22, 26]; as a consequence, the solution provided by an IA method for PF solution is not always as informative as expected. Thus in [27], we showed that the use of IA for the solution of power flow equations may easily yield aberrant solutions, due to the fact that the IA formalism is unable to represent the correlations that the power flow equations establishes between the power systems state variables; as a consequence, at each algorithm step spurious values are added to the solutions, which could converge to large domains that include the correct solution. This phenomenon is well known in the simulation of qualitative systems [28, 29], and requires the adoption of specific techniques such as the Interval Gauss elimination, the Krawczyk's method, and the Interval Gauss Seidel iteration procedure. Therefore, the application of these paradigms in the PF solution process leads to realistic solution bounds only for certain special classes of matrices (e.g. M-matrices, H-matrices, diagonally dominant matrices, tri-diagonal matrices) [30]; furthermore, to guarantee convergence, it is necessary to preconditioning the linear PF equations by an M-matrix [31]. These techniques make the application of IA to PF analysis complex and time consuming.

1.5 Affine Arithmetic-Based Methods

To overcome the aforementioned limitations in IA, in [27], we proposed the employment of a more effective self validated paradigm based on Affine Arithmetic (AA) to represent the uncertainties of the PF state variables, which is one of the topics of the present thesis. In this approach, each state variable can be expressed by a first degree polynomial composed by a central value, i.e. the nameplate value, and a number of partial deviations that represent the correlation among various variables. The adoption of AA for uncertainty representation allows expressing the power flow equations in a more convenient formalism, so that a reliable estimation of the PF solution hull can be computed taking into account the parameter uncertainty interdependencies, as well as the diversity of uncertainty sources. The main advantage of this solution strategy is that it requires neither derivative computations nor interval systems, being thus suitable in principle for large scale power flow studies, where robust and computationally efficient solution algorithms are required. These benefits have been confirmed in [32] and in [33], which allows to determine operating margins for thermal generators in systems with uncertain parameters, by representing all the state and control variables with affine forms accounting for forecast, model error, and other sources of uncertainty, without the need to assume a probability density function. These methodologies have been recently recognized as one of the most promising alternative for stochastic information management in bulk generation and transmission systems for smart grids [34].

Based on our own work reported in [27], several papers have explored the application of AA-based computing in power system analysis. In particular, in [35] the state estimation problem in the presence of mixed phasor and conventional power measurements has been addressed, considering the effect of network parameters uncertainty by an iterative weight least square algorithm based on IA and AA processing. In [36], an AA-based model of the uncertain OPF problem is proposed, using complementarity conditions to properly represent generator bus voltage controls, including reactive power limits and voltage recovery; the model is then used to obtain operational intervals for the PF variables considering active and reactive power demand uncertainties. In [37], a non-iterative solution scheme based on AA is proposed to estimate the bounds of the uncertain PF solutions by solving an uncertain PF problem, which is formalized by an interval power flow problem and solved by quadratic programming optimization models.

The benefits deriving from the application of AA-based computing to power system planning and operation in the presence of data uncertainty have been assessed in [38], which confirms that AA represents a fast and reliable computing paradigm that allows planners and operators to cope with high levels of renewable energy penetration, electric vehicle load integration, and other uncertain sources. Moreover, as confirmed in [32, 33, 39, 40], AA allows the analyst to narrow the gap between the upper and lower bounds of the OPF solutions, avoiding the overestimation of bounds resulting from correlation of variables in IA.

Although the aforementioned papers offer considerable insight on the role that AA may play in power systems analysis, several open problems still remain unsolved, particularly:

- Further exploration of the application of AA-based techniques to uncertain OPF analysis.
- Rigorous methodologies aimed at selecting the noise symbols of the affine forms representing the power system state variables.
- More efficient paradigms aimed at reducing the overestimation errors of AA-based PF and OPF problems.

2 Chapter Contributions

Based on the above literature review, the following are the main Chapter objectives:

1. Demonstrate with a realistic test system that the use IA in PF and OPF analysis leads to over-pessimistic estimation of the solution hull, which are not useful in most practical applications due to the inability of IA to keep track of correlations between the power systems state variables, and analyze the employment of AA to represent the uncertainties of the power systems state variables. The adoption of AA for uncertainty representation will allow to express the OPF models in a more convenient formalism compared to the traditional and widely used linearization frequently used in interval Newton methods.

2. Present and thoroughly test a novel solution methodology based on AA for OPF studies with data uncertainties. By using the proposed methodology, a reliable estimation of the OPF solutions hull will be computed, taking into account the parameter uncertainty inter-dependencies as well as the diversity of uncertainty sources. The main advantage of this solution strategy is that it doe not require the solution of interval systems of equations, being thus suitable in principle for large scale OPF studies where robust and computationally efficient solution algorithms are required.

References

1. Verbic, G., Cañizares, C.A.: Probabilistic optimal power flow in electricity markets based on a two-point estimate method. IEEE Trans. Power Syst. **21**(4), 1883–1893 (2006)
2. Wan, Y.H., Parsons, B.K.: Factors Relevant to Utility Integration of Intermittent Renewable Technologies. National Renewable Energy Laboratory (1993)
3. Chen, P., Chen, Z., Bak-Jensen, B.: Probabilistic load flow: a review. In: Proceedings of the 3rd International Conference on Deregulation and Restructuring and Power Technologies, DRPT 2008, pp. 1586–1591 (2008)
4. Zou, B., Xiao, Q.: Solving probabilistic optimal power flow problem using quasi Monte Carlo method and ninth-order polynomial normal transformation. IEEE Trans. Power Syst. **29**(1), 300–306 (2014)
5. Hajian, M., Rosehart, W.D., Zareipour, H.: Probabilistic power flow by monte carlo simulation with latin supercube sampling. IEEE Trans. Power Syst. **28**(2), 1550–1559 (2013)
6. Zhang, H., Li, P.: Probabilistic analysis for optimal power flow under uncertainty. IET Gener. Transm. Distrib. **4**(5), 553–561 (2010)
7. Yu, H., Chung, C.Y., Wong, K.P., Lee, H.W., Zhang, J.H.: Probabilistic load flow evaluation with hybrid latin hypercube sampling and cholesky decomposition. IEEE Trans. Power Syst. **24**(2), 661–667 (2009)
8. Mori, H., Jiang, W.: A new probabilistic load flow method using mcmc in consideration of nodal load correlation. In: Proceedings of the 15th International Conference on Intelligent System Applications to Power Systems, pp. 1–6 (2009)
9. Yu, H., Rosehart, B.: Probabilistic power flow considering wind speed correlation of wind farms. In: Proceedings of the 17th Power Systems Computation Conference, pp. 1–7 (2011)
10. Zhang, H., Li, P.: Probabilistic power flow by monte carlo simulation with latin supercube sampling. IET Gener., Transm. Distrib. **4**(5), 553–561 (2010)
11. Schellenberg, A., Rosehart, W., Aguado, J.: Cumulant-based probabilistic optimal power flow (p-opf) with gaussian and gamma distributions. IEEE Trans. Power Syst. **20**(2), 773–781 (2005)
12. Meliopoulos, A.P.S., Cokkinides, G.J., Chao, X.Y.: A new probabilistic power flow analysis method. IEEE Trans. Power Syst. **5**(1), 182–190 (1990)
13. Allan, R.N., da Silva, A.M.L., Burchett, R.C.: Evaluation methods and accuracy in probabilistic load flow solutions. IEEE Trans. Power Appar. Syst., PAS **100**(5), 2539–2546 (1981)
14. Zhang, P., Lee, S.T.: Probabilistic load flow computation using the method of combined cumulants and gram-charlier expansion. IEEE Trans. Power Syst. **19**(1), 676–682 (2004)
15. Sanabria, L.A., Dillon, T.S.: Stochastic power flow using cumulants and von mises functions. Int. J. Electr. Power Energy Syst. **8**(1), 47–60 (1986)
16. Dimitrovski, A., Tomsovic, K.: Boundary load flow solutions. IEEE Trans. Power Syst. **19**(1), 348–355 (2004)
17. Alvarado, F., Hu, Y., Adapa, R.: Uncertainty in power system modeling and computation. In: Proceedings of the IEEE International Conference on Systems, Man and Cybernetics, pp. 754–760 (1992)

18. Vaccaro, A., Villacci, D.: Radial power flow tolerance analysis by interval constraint propagation. IEEE Trans. Power Syst. **24**(1), 28–39 (2009)
19. Madrigal, M., Ponnambalam, K., Quintana, V.H.: Probabilistic optimal power flow. In: Proceedings of the IEEE Canadian Conference on Electrical and Computer Engineering, vol. 1, pp. 385–388 (1998)
20. Li, X., Li, Y., Zhang, S.: Analysis of probabilistic optimal power flow taking account of the variation of load power. IEEE Trans. Power Syst. **23**(3), 992–999 (2008)
21. Mohammadi, M., Shayegani, A., Adaminejad, H.: A new approach of point estimate method for probabilistic load flow. Int. J. Electr. Power Energy Syst. **51**, 54–60 (2013)
22. Stolfi, J., De Figueiredo, L.H.: Self-validated numerical methods and applications. In: Proceedings of the Monograph for 21st Brazilian Mathematics Colloquium. Citeseer (1997)
23. Moore, R.: Methods and Applications of Interval Analysis, vol. 2. SIAM (1979)
24. Wang, S., Xu, Q., Zhang, G., Yu, L.: Modeling of wind speed uncertainty and interval power flow analysis for wind farms. Autom. Electr. Power Syst. **33**(1), 82–86 (2009)
25. Pereira, L.E.S., Da Costa, V.M., Rosa, A.L.S.: Interval arithmetic in current injection power flow analysis. Int. J. Electr. Power Energy Syst. **43**(1), 1106–1113 (2012)
26. Neher, M.: From interval analysis to Taylor models-an overview. In: Proceedings of the International Association for Mathematics and Computers in Simulation (2005)
27. Vaccaro, A., Cañizares, C.A., Villacci, D.: An affine arithmetic-based methodology for reliable power flow analysis in the presence of data uncertainty. IEEE Trans. Power Syst. **25**(2), 624–632 (2010)
28. Armengol, J., Travé-Massuyès, L., Vehi, J., de la Rosa, J.L.: A survey on interval model simulators and their properties related to fault detection. Annu. Rev. Control. **24**, 31–39 (2000)
29. Bontempi, G., Vaccaro, A., Villacci, D.: Power cables' thermal protection by interval simulation of imprecise dynamical systems. IEE Proc.-Gener., Transm. Distrib. **151**(6), 673–680 (2004)
30. Barboza, L.V., Dimuro, G.P., Reiser, R.H.S.: Towards interval analysis of the load uncertainty in power electric systems. In: Proceedings of the International Conference on Probabilistic Methods Applied to Power Systems, pp. 538–544 (2004)
31. Alvarado, F., Wang, Z.: Direct Sparse Interval Hull Computations for Thin Non-M Matrices (1993)
32. Vaccaro, A., Cañizares, C.A., Bhattacharya, K.: A range arithmetic-based optimization model for power flow analysis under interval uncertainty. IEEE Trans. Power Syst. **28**(2), 1179–1186 (2013)
33. Pirnia, M., Cañizares, C.A., Bhattacharya, K., Vaccaro, A.: An affine arithmetic method to solve the stochastic power flow problem based on a mixed complementarity formulation. Electr. Power Compon. Syst. **29**(6), 2775–2783 (2014)
34. Hao, L., Tamang, A.K., Weihua, Z., Shen, X.S.: Stochastic information management in smart grid. IEEE Commun. Surv. Tutor. **16**(3), 1746–1770 (2014)
35. Rakpenthai, C., Uatrongjit, S., Premrudeepreechacharn, S.: State estimation of power system considering network parameter uncertainty based on parametric interval linear systems. IEEE Trans. Power Syst. **27**(1), 305–313 (2012)
36. Pirnia, M., Cañizares, C.A., Bhattacharya, K., Vaccaro, A.: A novel affine arithmetic method to solve optimal power flow problems with uncertainties. In: Proceedings of the IEEE Power and Energy Society General Meeting, pp. 1–7 (2012)
37. Bo, R., Guo, Q., Sun, H., Wenchuan, W., Zhang, B.: A non-iterative affine arithmetic methodology for interval power flow analysis of transmission network. Proc. Chin. Soc. Electr. Eng. **33**(19), 76–83 (2013)
38. Wang, S., Han, L., Zhang, P.: Affine arithmetic-based dc power flow for automatic contingency selection with consideration of load and generation uncertainties. Electr. Power Compon. Syst. **42**(8), 852–860 (2014)
39. Wei, G., Lizi, L., Tao, D., Xiaoli, M., Wanxing, S.: An affine arithmetic-based algorithm for radial distribution system power flow with uncertainties. Int. J. Electr. Power Energy Syst. **58**, 242–245 (2014)
40. Ding, T., Cui, H.Z., Gu, W., Wan, Q.L.: An uncertainty power flow algorithm based on interval and affine arithmetic. Autom. Electr. Power Syst. **36**(13), 51–55 (2012)

Balancing Waste Water Treatment Plant Load Using Branch and Bound

Ronald van Nooijen and Alla Kolechkina

Abstract The problem of smoothing dry weather inflow variations for a Waste Water Treatment Plant (WWTP) that receives sewage from multiple mixed sewer systems is formulated. A first rough control algorithm that uses branch and bound is presented. The control algorithm uses a form of Model Predictive Control. Trials showed that the algorithm had trouble satisfying two constraints that were initially regarded as 'soft' constraints. As a result, a closer look was taken at the feasibility of the problem. A family of simpler problems were derived to do so. These auxiliary problems made it possible to show that for a given subclass feasibility was strongly dependent on the choice of problem parameters.

1 Introduction

In most of the Netherlands as in other reactively flat regions, municipal sewer networks consist of separate sub-networks where gravity driven flow transports the sewage. These sub-networks are linked by pump stations [4, 5]. A sub-network serves to collect sewage, to transport locally collected sewage to the pump station, and may also transport sewage from an upstream sub-network to a downstream sub-network. For each municipality there are also one or more pump stations that transport the sewage from the network to a Waste Water Treatment Plant (WWTP) through a pressurized pipeline. In a 'combined' sewer system, the sewer functions both as a foul water sewer and a storm drain. In these systems the pipes are over-dimensioned

R. van Nooijen (✉)
Faculty of Civil Engineering and Geosciences, Delft University of Technology,
Stevinweg 1, 2628 Delft, CN, The Netherlands
e-mail: r.r.p.vannnooyen@tudelft.nl

A. Kolechkina
Delft Center for Systems and Control, Faculty of Mechanical, Maritime and Materials
Engineering, Delft University of Technology, Mekelweg 2, 2628 CD Delft, The Netherlands
e-mail: a.g.kolechkina@tudelft.nl

© Springer Nature Switzerland AG 2020   197
M. Ceberio and V. Kreinovich (eds.), *Decision Making under Constraints*,
Studies in Systems, Decision and Control 276,
https://doi.org/10.1007/978-3-030-40814-5_25

to allow for the wet weather flow. In an 'improved separated' sewer system part of the flow into the storm drains is diverted into the foul water sewer to deal with the 'first flush', the sewage at the start of a precipitation event that contains a relatively high proportion of pollutants.

Each pump station has a wet well to even out the supply to the pumps and avoid air ingestion. A pump station usually has multiple pumps, for instance, one pump for use during dry weather and a bigger pump for use during wet weather. Some of these pumps are operated on an on/off basis; others can run at a range of different speeds. The latter type usually has a lower and an upper limit on the speed and hence the discharge. A pump station with multiple pumps and different discharge ranges may have both a discrete state, that records which pumps are on and off, and a continuous state, the current pump discharges.

It is usual to impose limits on the number of on/off cycles per hour as starting a pump causes additional wear and tear on the equipment, which leads to higher maintenance cost and shorter pump life.

In 2015 representatives of two water boards and one municipality decided to look into a long standing problem. The Garmerwolde WWTP receives water from several pressurized pipelines. These in turn receive sewage from 5 local sewer networks through pumping stations. The WWTP was designed to process sewage as it arrives in the plant; there are no buffers. During dry weather the supply of sewage to the pump stations varied roughly sinusoidally over a 24 h period, and the 5 pump stations were under local control (no coordination). This led to extremely uneven supply to the WWTP, which in turn led to high costs for chemical additives and air injection. Moreover, at times three large pumping stations could be using the same pressurized line, which wasted energy as well. At dry weather flow rates there is considerable room to store the flow in the sewer networks, moreover, the dry weather flows are considered to be much less unpredictable than wet weather flows. It was therefore proposed to examine the possibility to coordinate the use of the pumping stations during dry weather with the aim of realizing an approximately constant flow to the WWTP.

The only way to realize a more even flow to the WWTP was to use the local sewer networks as temporary storage. However, the sewage in local sewer systems should not be stationary for long periods to avoid silting up of the pipes. Another challenge was the timely detection of wet weather, as the storage in use for evening out the load on the WWTP would then be needed to store peak intensity rain fall that the pumps could not immediately cope with. More information on the design and operation of Dutch combined sewer systems can be found in NLingenieurs Sewer Systems Workgroup [2].

During the project it was found that is surprisingly easy to end up with a problem that has no solution. In this study we will formulate a dynamical model that covers the situation in Garmerwolde, specify constraints on the pump station settings corresponding to the desired evening out of the flow, describe a simple algorithm used for pump control in the testing phase, and determine some quick tests on the constraints to detect an unsolvable problem. Earlier work on the existence of solutions can be found in van Nooijen and Kolechkina [6].

2 Model Formulation

The local sewer networks will be modelled as reservoirs. We have m reservoirs and an n time step inflow forecast for each reservoir. The control time step length is Δt; the start of time step k will be referred to as t_k. For each reservoir there is a time dependent regulated outflow. We will use lower case letters for scalars or vectors and upper case letters for closed finite intervals.

A pump station may have multiple pumps and may be able to use different combinations of pumps, moreover, the availability of these combinations may be time dependent, but in the Garmerwolde project a station was treated as if it had just one pump with state $q_i(t) \in 0 \cup Q_i$, where $Q_i = \left[\underline{q}_i, \bar{q}_i\right]$ and $\underline{q}_i > 0$. While the sewer network upstream of each pumping station has a quite complex shape, it was modelled as one volume with state $v_i(t) \in V_i$ where $V_i = \left[\underline{v}_i, \bar{v}_i\right]$ with $\underline{v}_i \geq 0$ gives the range of storage volumes that the automatic control system is allowed to use. In real systems there would be some "dead storage", a volume in sewer network i below the lowest water level at which the pumps can safely operate, but dropping below this level would usually trigger an alert and an emergency pump stop, so in the model $v_i(t)$ represents the storage volume above this water level. In the model we will assume that a local controller will switch off the pump when the level drops to $\underline{v}_i$, but will then hand the pump back to central control at the start of the next time step. There would usually also be a water level above $\bar{v}_i$ at which a station would go to maximum capacity to avoid spills. In the model we will assume that a local controller will switch the pump to maximum flow $\bar{q}_i$ when the level rises to $\bar{v}_i$, but will then hand the pump back to central control at the start of the next time step.

Finally it is necessary to empty the system periodically to avoid excessive sedimentation. As it may not be necessary (or even possible) to drop to $\underline{v}_i$, the level $v_{e,i}$ is introduced to represent a system volume that when reached is low enough for the purposes of excessive sedimentation prevention.

The flow presented to pumping station i by its sewer network will be represented in the model by an inflow $q_i^{in}(t)$ into reservoir i at time t; this is a non-negative real number. We assume that the following is known about the inflow for all t and i

$$q_i^{in}(t) \in Q_i^{in} = \left[\underline{q}_i^{in}, \bar{q}_i^{in}\right] \tag{1}$$

and there is a $C_i^{in}(\tau)$ such that for all $\tau \geq 0$ and all t and i

$$\int_{u=t}^{t+\tau} q_i^{in}(u)\,du \in \tau \times Q_i^{in} \times C_i^{in}(\tau) \tag{2}$$

with $C_i^{in}(\tau) = \left[\underline{c}_i^{in}(\tau), \bar{c}_i^{in}(\tau)\right]$ such that $\underline{c}_i^{in}(\tau) \geq 0$ and $\bar{c}_i^{in}(\tau) \leq 1$.

The controller will supplied with $q_i^{fc}(k)$, a forecast of the average inflow into reservoir i for time step k

$$q_i^{\text{av}}(k) = \frac{1}{\Delta t} \int\limits_{t=t_k}^{t_{k+1}} q_i^{\text{in}}(t)\, dt \tag{3}$$

The commands from the central controller to the pump stations will be represented by the selected discharge $q_i(k)$ for time step k.

To summarize: we will model the storage in the network as a volume, and consider the storage and the pump station together as one system with inflows $q_i^{\text{in}}(t)$ and commands $q_i(k)$ as input and $v_i(t)$ and $q_i(t)$ as state.

To formulate the state evolution in time, we introduce the following auxiliary functions

$$t_k \le t \le t_{k+1} : \tilde{v}_i(t;k) = v_i(t_k) + \int\limits_{u=t_k}^{t} q_i^{\text{in}}(u) - q_i(k)\, du \tag{4}$$

with $\tilde{v}_i(t;k)$ continuous on $\left[t_k, t_{k+1}\right]$,

$$\tilde{t}_{\text{off}}(t;k) = \begin{cases} \inf\left\{t_k \le t \le t_{k+1} : \tilde{v}_i(t;k) \le v_{\min,i}\right\} & q_i(k) > 0 \\ t_k & q_i(k) = 0 \end{cases} \tag{5}$$

$$\tilde{t}_{\text{on}}(t;k) = \begin{cases} \inf\left\{t_k \le t \le t_{k+1} : \tilde{v}_i(t;k) \ge v_{\max,i}\right\} & q_i(k) = 0 \\ t_k & q_i(k) > 0 \end{cases} \tag{6}$$

where we used the convention that for $\emptyset \subset \mathbb{R}$ we set $\inf \emptyset = \infty$. To simplify the formulation of the time evolution, we assume that

$$\frac{\bar{v}_i - \underline{v}_i}{\bar{q}_i} > \Delta t \tag{7}$$

and

$$\frac{\bar{v}_i - \underline{v}_i}{\bar{q}_i^{\text{in}}} > \Delta t \tag{8}$$

to avoid two local controller interventions within the same time step. We now define the time evolution of $q_i(t)$ by

$$t_k \le t < t_{k+1} : q_i(t) = \begin{cases} 0 & q_i(k) = 0,\ t < \tilde{t}_{\text{on}}(t;k) \\ \bar{q}_i & q_i(k) = 0,\ t \ge \tilde{t}_{\text{on}}(t;k) \\ q_i(k) & q_i(k) > 0,\ t < \tilde{t}_{\text{on}}(t;k),\ t < \tilde{t}_{\text{off}}(t;k) \\ 0 & q_i(k) > 0,\ t < \tilde{t}_{\text{on}}(t;k),\ t \ge \tilde{t}_{\text{off}}(t;k) \\ \bar{q}_i & q_i(k) > 0,\ t \ge \tilde{t}_{\text{on}}(t;k) \end{cases} \tag{9}$$

Note that $q_i(t)$ is piecewise constant and right continuous. This $q_i(t)$ can then be used to define the time evolution of $v_i(t)$ by

$$v_i(t) = v_i(t_k) + \int_{t'=t_k}^{t} q_{in,i}(t') - q_i(t') \, dt' \tag{10}$$

Based on the state the following auxiliary functions are defined, the most recent time the pump was switched off

$$t_i^{off}(t) = \sup\left\{t' \leq t : q_i(t') = 0, \; q_i(t'^-) > 0\right\} \tag{11}$$

the most recent time the pump was switched on

$$t_i^{on}(t) = \sup\left\{t' \leq t : q_i(t') > 0, \; q_i(t'^-) = 0\right\} \tag{12}$$

and the most recent time the reservoir was considered empty

$$t_{e,i}(t) = \sup\left\{t' \leq t : v_i(t') \leq v_{e,i}\right\} \tag{13}$$

The following constraints will be imposed on the system state:

$$q_i(t) \in Q_i \tag{14}$$

$$v_i(t) \in V_i \tag{15}$$

and

$$\sum_{i=1}^{m} q_i(t) \in Q_{tgt} \tag{16}$$

where the interval $Q_{tgt} = \left[\underline{q}_{tgt}, \overline{q}_{tgt}\right]$ represents the range of allowable flow rates into the WWTP. The following additional constraints may be imposed:

- An upper limit on the time $\tau_{e,i}$ between successive moments of dropping to $v_{e,i}$

$$t - t_{e,i}(t) \leq \tau_{e,i} \tag{17}$$

- A lower limit $\tau_i^{off} > 0$ on the time between switching the pump off and on

$$q_i(t) > 0 \Rightarrow t_i^{on}(t) - t_i^{off}(t) < \tau_i^{off} \tag{18}$$

- A lower limit $\tau_i^{on} > 0$ on the time between switching the pump on and off

$$q_i = 0 \Rightarrow t_i^{off}(t) - t_i^{on}(t) < \tau_i^{on} \tag{19}$$

3   General Problem to Be Solved for Known Inflow

The problem to be solved can now be formulated as follows.

Problem 1 Given an initial state $v_i(t_0) = v_{0,i} \in V_i$, $q_i(t_0^-) = q_{0,i} \in Q_i$ for $i = 1, 2, \ldots, m$ and

$$\sum_{i=1}^{m} q_{0,i} \in Q_{\text{tgt}} \tag{20}$$

determine $q_i(k)$ for $i = 1, 2, \ldots, m$, $k = 0, 1, \ldots, n$ and $t_0 \leq t < t_n$ such that

$$q_i(k) \in Q_i \tag{21}$$

$$\sum_{i=1}^{m} q_i(t) \in Q_{\text{tgt}} \tag{22}$$

$$v_i(t) \in V_i \tag{23}$$

So even without the time based constraints, we already have $2 \times (2m + 1)$ inequalities per time step and m variables (the discharges) per time step. These variables might be either integer, continuous or continuous with gaps in the allowed value range. In addition, we have the soft constraints on run times and on emptying the system. For a simplified problem without run time or emptying constraints with 5 on/off pumps, a 15 min time step and 24 h look-ahead, we get 2112 inequalities for 480 binary variables.

Lemma 1 *If Problem 1 has a solution then for $i = 1, 2, \ldots, m$ and $k = 0, 1, \ldots, n - 1$ then we can define*

$$q_{i,k} = \max \{q_i(t) : t_k < t < t_{k+1}\} \tag{24}$$

$$t_{i,k} = \mu(\{t : t_k \leq t < t_{k+1}, q_i(t) > 0\}) \tag{25}$$

where μ is the standard Lebesgue measure on the real line. Moreover, there is an integer N and a finite sequence of times $\hat{t}_u$ with $u = 1, 2, \ldots, N$ such that $\hat{t}_0 = t_0$, $\hat{t}_N = t_n$ and for $1 \leq u < N$ at time $\hat{t}_u$ at least one $q_i(t)$ has a discontinuity, and a $q_i(t)$ has discontinuities only at times in the sequence $\hat{t}_u$.

Proof The quantity in (24) is well defined because $q_i(t)$ is piecewise constant. The quantity in (25) is well defined and corresponds to the duration of non-zero pump discharge because $q_i(t)$ is piecewise constant. As a pump i can start or stop only at one of the times t_k or at most once inside an interval $[t_k, t_{k+1}]$ thanks to conditions (7) and (8) it follows that: if $t_{i,k} < \Delta t$ then either $q_i(t) > 0$ for $t_k \leq t < t_k + t_{i,k}$ or $q_i(t) > 0$ for $t_{k+1} - t_{i,k} \leq t < t_{k+1}$. There are at most $3m(n + 1)$ discontinuities in the vector with components $q_i(t)$ so $N \leq 3m(n + 1)$. Now build $\hat{t}_u$ as follows.

Set $\hat{t}_u = t_0$. Now repeat: find the first time $\tilde{t} > \hat{t}_u$ where any $q_i(t)$ is discontinuous, set $\hat{t}_{u+1} = \tilde{t}$ and set $u + 1 = u$. If t_n has not been added after exhausting all vector discontinuities, set $\hat{t}_{u+1} = t_n$. □

From this lemma it follows that from any solution of Problem 1 we can construct a solution of Problem 2.

Problem 2 Given an initial state $v_i(t_0) = v_{0,i} \in V_i$, $q_i(t_0^-) = q_{0,i} \in Q_i$ for $i = 1, 2, \ldots, m$ and

$$\sum_{i=1}^{m} q_{0,i} \in Q_{\text{tgt}} \tag{26}$$

determine $q_{i,u}$ for $i = 1, 2, \ldots, m$, $u = 0, 1, \ldots, N - 1$ and $\hat{t}_0 \leq t < \hat{t}_N$ such that

$$q_{i,u} \in Q_i \tag{27}$$

$$\sum_{i=1}^{m} q_{i,u} \in Q_{\text{tgt}} \tag{28}$$

$$v_i(t) \in V_i \tag{29}$$

where for $\hat{t}_u \leq t < \hat{t}_{u+1}$

$$v_i(t) = v_i(\hat{t}_u) + \int_{t'=\hat{t}_u}^{t} q_i^{\text{in}}(t') \, dt' - (t - \hat{t}_u) \sum_{i=1}^{m} q_{i,u}$$

The constraint (17) can again be imposed.

Lemma 2 *If Problem 2 with $\tau_{e,i} = n_e \Delta t$ in constraint (17) and $n > 2n_e$ and $q_i^{\text{in}}(t')$ periodic with period n_e has a solution, then for all i*

$$\int_{t=t_0}^{t_{n_e}} q_i^{\text{in}}(t) \, dt \leq n_e \Delta t \min(\bar{q}_i, \bar{q}_{\text{tgt}}) \tag{30}$$

Proof (Sketch) Equation 30 must hold else it would not be possible to have and $n > 2n_e$ and $t_{e,i}(t) \leq n_e \Delta t$ at all t. □

Lemma 3 *If Problem 2 with $\tau_{e,i} = n_e \Delta t$ in constraint (17) and $n = \infty$ and $q_i^{\text{in}}(t')$ periodic with period n_e has a solution, then*

$$\sum_{i=1}^{m} \int_{t=t_0}^{t_{n_e}} q_i^{\text{in}}(t) \, dt \leq n_e \Delta t \bar{q}_{\text{tgt}} \tag{31}$$

Proof (Sketch) If (31) does not hold then the total volume in the system will go to infinity as t goes to infinity. This means that for at least one i, the value of $v_i(t)$ will go to infinity. As the pump capacity per period is limited to $n_e \Delta t \min(\bar{q}_i, \bar{q}_{tgt})$, this means (17) will eventually fail for this i.   ◻

4 Typical Input Data

In the Garmerwolde case there are five pumping stations. The inflow exhibits a roughly periodic sinusoidal pattern with a length of 24 h with a minimum around 6 a.m. Time step length is 15 min. Minima and maxima for the inflow patterns are given in Table 1.

Note that the dry weather hourly inflows for all but Groningen are always below the minimum pumping capacity given for the station in Table 2. Two examples for the V and Q intervals are given in Table 2. A typical target flow range would be $Q_{tgt} = [1700, 2200]$.

Even with only $O(10)$ possible combinations of pumps per time step, the number of possible selections in the unconstrained problem with 15 min time step and 24 h look ahead is $O(10^{96})$. For the problem with variable flows there is the added complexity of selecting the flow for each pump. If the pumps were simple on/off pumps, then fitting them into the range of target flows would be remarkably like the two dimensional cutting stock problem Dyckhoff [1]. The problem also has points in

Table 1 Minima and maxima of hourly inflow

Pump station	Inflow	
	min (m³/h)	max (m³/h)
Groningen (GR)	454	1179
Selwerd (SE)	126	423
G. Huizinga (GH)	144	600
Haren W. (HW)	80	155
Lewenborg (LE)	2	335

Table 2 Example of data used in program

Pump station	Volumes $v_{e,i}$, $[\underline{v}, \bar{v}]$ (m³)		Pump capacity (m³/h)	
	Example 1	Example 2	Example 1	Example 2
Groningen (GR)	1000, [95, 3957]	64, [26, 14380]	[1800, 2200]	[1500, 2000]
Selwerd (SE)	121, [38, 296]	116, [0, 4328]	[1620, 1980]	[1060, 1350]
G. Huizinga (GH)	176, [150, 300]	176, [134, 9381]	[330, 400]	[1600, 1850]
Haren W. (HW)	27, [22, 50]	27, [0, 1097]	[1020, 1320]	[600, 650]
Lewenborg (LE)	30, [9, 239]	30, [0, 8969]	[1440, 1760]	[550, 1900]

common with the problem addressed in Perkins and Kumark [3], but that approach does not allow for variable part supply rates. A result on the existence of solutions for a very special case is discussed in van Nooijen and Kolechkina [6].

5   A Simple Greedy Algorithm

The algorithm presented here is intended to be used in the manner of Model Predictive Control (MPC). This means the algorithm is run to find a promising sequence of control actions up to a forecast horizon and that only the first of those control actions is executed. For the next control step this process is repeated.

The algorithm does a depth first search for a path of length n in a tree where the nodes correspond to system states at times t_k and the edges correspond to choices of flow ranges. At each node it generates all possible combinations of pump flow states (branch), filters out those that are certain to lead to volume or target flow constraint violations (bound), orders the remainder so that it will first try those that respect both 'minimum' run time and 'time since empty' constraints, then those that respect only 'minimum run time', followed by those that respect only 'time since empty' and, finally, those that do not respect either of these constraints. Within each group, the flow state combinations where the pumps for districts furthest from empty are active, are taken first. If a path turns out to be a dead end, the algorithm backtracks upward. The first path to reach the horizon is used. For the selection of flow ranges in the first step along the path specific discharges are calculated. The main purpose of the search is to cope with the periodic variation in the inflow. Once a path is found, the selection of flow ranges for the first step needs to be translated into actual pump flows.

5.1   A Sketch of a Greedy Algorithm with Interval Arithmetic

The pumps can deliver a range of discharges and the WWTP can accept a range of discharges, if the ranges are not too wide, interval arithmetic can be used to check whole ranges in stead of individual values. For narrow pump ranges Q and a narrow target range Q^{tgt}, it makes sense to work in interval arithmetic for the system state. For wide ranges, further investigations are needed.

We proceed as follows

$$\tilde{V}_i(t_0) = v_i\left(t_0^-\right) \tag{32}$$

A time step is processed as follows. For each $\mathbf{j} \in \{0, 1\}^m$ we apply the following accept/reject process. We calculate

$$Q_i''(k) = \begin{cases} \left(\frac{V_i - (\tilde{V}_i(t_k) + q_i^{fc}(k)\Delta t)}{\Delta t} \right) \cap \{0\} & j_i = 0 \\ \left(\frac{V_i - (\tilde{V}_i(t_k) + q_i^{fc}(k)\Delta t)}{\Delta t} \right) \cap Q_{j_i} & j_i = 1 \end{cases} \tag{33}$$

If any $Q_i''(k)$ is empty, then reject this $\mathbf{j}$, else determine

$$Q_i'(k) = \begin{cases} Q_i''(k) & j_i = 0 \\ Q_i''(k) \cap \left(Q_{tgt} - \sum_{u=1, u \neq i}^m Q_{j_u} \right) & j_i = 1 \end{cases} \tag{34}$$

If any $Q_i'(k)$ is empty, then reject this $\mathbf{j}$, else do the tests

1. $\left(\sum_{i=1}^m Q_i'(k) \right) \cap Q_{tgt} \neq \emptyset$
2. $\left(\tilde{V}_i(t_k) + \Delta t \left(q_i^{fc}(k) - Q_i'(k) \right) \right) \cap V_i \neq \emptyset$

If both tests succeed then we define

$$\tilde{V}_i(t) = \left(\tilde{V}_i(t_k) + (t - t_k) \left(q_i^{fc}(k) - Q_i'(k) \right) \right) \tag{35}$$

$$\tilde{V}_i(t_{k+1}) = \left(\tilde{V}_i(t_k) + \Delta t \left(q_i^{fc}(k) - Q_i'(k) \right) \right) \cap V_i(k+1) \tag{36}$$

Next we define

$$t_k \leq t < t_{k+1} : Q_i'(t) = Q_i'(k)$$

$$t_i^{off}(t) = \sup \left\{ t' \leq t : Q_i'(t') = \{0\}, \ Q_i'(t'^-) > 0 \right\} \tag{37}$$

$$t_i^{on}(t) = \sup \left\{ t' \leq t : Q_i'(t') > 0, \ Q_i'(t'^-) = \{0\} \right\} \tag{38}$$

$$t_e(t) = \begin{cases} 0 & v_{e,i} \in \tilde{V}_i(t) \\ t - t_k & v_{e,i} \in \tilde{V}_i(t_k) \text{ and } \tilde{V}_i(t) > v_{e,i} \\ t_e(t_k) + (t - t_k) & \tilde{V}_i(t_k) > v_{e,i} \text{ and } \tilde{V}_i(t) > v_{e,i} \end{cases} \tag{39}$$

If a path is found, the widening due to interval arithmetic may mean it does not contain an actual path with specific point values for the discharges. However, the forecast is not that accurate, so the widening might help as well as hinder. Moreover, the process is repeated every control time step, so errors can be corrected.

Once a path is found, a specific allocation of pump discharges is derived for one time step. Several versions of such an allocation are being tested.

6   Checks for the Feasibility of Constraints

The formulation of conditions for the existence of a solution for Problem 1 with just constraint (17) on time between empty reservoirs proved surprisingly hard. It turned out to be quite easy to formulate infeasible problems. It was therefore decided to look for quick tests for infeasible constraints. From numerical experiments it was clear that one of the sources of the problem was the forced cooperation between different pumping stations to meet the constraint on total flow to the WWTP (16) in combination with the need to reach empty status for all the reservoirs.

A simpler problem was formulated as well as a conjecture about its relation to Problem 1.

6.1   A Simpler Problem

Problem 3 Suppose we have m-dimensional vectors v_0 and v_e of real non-negative numbers; an m-dimensional vector Q with real non-empty, non-negative closed intervals; a real non-empty, non-negative closed interval Q_{tgt}; a given fixed j and the set

$$\mathcal{G}_j = \left\{ G : j \in G \subseteq \{1, 2, \ldots, m\}, \left(\sum_{i \in G} Q_i \right) \cap Q_{tgt} \neq \emptyset \right\} \qquad (40)$$

Let $K = |\mathcal{G}_j|$, the number of elements of $\mathcal{G}_j$. Find an indexing function $I : \{1, 2, \ldots, K\} \to \mathcal{G}_j$, a finite sequence t_k, and numbers $q_{i,k} \in Q_i$ such that

$$q_{i,k} \in \begin{cases} \{0\} & i \notin G_k \\ Q_i & i \in G_k \end{cases}$$

$$k = 1, 2, \ldots, K : \sum_{i \in G} q_{i,k} \in Q_{tgt} \qquad (41)$$

$$i = 1, 2, \ldots, m : v_{0,i} - \sum_{k=1}^{K} q_{i,k} t_k \geq \underline{v}_i \qquad (42)$$

$$\underline{v}_j \leq v_{0,j} - \sum_{k=1}^{K} q_{j,k} t_k \leq v_{e,j} \qquad (43)$$

where $G_k = I(k)$.

Conjecture 1 If Problem 2 with $\tau_{e,i} = n_e \Delta t$ in constraint (17) can be solved for a $q_i^{in}(t)$ that is periodic with period n_e for $n = \infty$, initial condition $i = 1, 2, \ldots, m :$ $v_{0,i} = \tilde{v}_{0,i}$ and inflows $q_{in,i}(t)$, then Problem 3 can be solved for

$$i = 1, 2, \ldots, m : v_{0,i} = v_{e,i} + \int_{t=t_0}^{t_{ne}} q_{in,i}(t)\, dt \qquad (44)$$

Problem 3 can in turn be reduced to Problem 4.

Problem 4 Suppose we have m-dimensional vectors v_0 and v_e of real non-negative numbers; an m-dimensional vector Q with real non-empty, non-negative closed intervals; a real non-empty, non-negative closed interval Q_{tgt}; a given fixed j and the set G_j as in (40). Find an indexing function $I : \{1, 2, \ldots, K\} \to G_j$, a finite sequence t_k, and numbers $q_{i,k} \in Q_i$ such that

$$q_{i,k} \in \begin{cases} \{0\} & i \notin G_k \\ Q_i & i \in G_k \end{cases}$$

$$q_{j,k} = \underline{q}_j + \min\left(\bar{q}_j - \underline{q}_j, \bar{q}_{tgt} - \sum_{i \in G_k} \underline{q}_i\right) \qquad (45)$$

$$k = 1, 2, \ldots, K : \sum_{i=1,i\neq j}^{m} q_{i,k} \in Q_{tgt} - q_{j,k} \qquad (46)$$

$$i = 1, 2, \ldots, m, i \neq j : v_{0,i} - \sum_{k=1}^{K} q_{i,k} t_k \geq \underline{v}_i \qquad (47)$$

$$\underline{v}_i \leq v_{0,i} - \sum_{k=1}^{K} q_{j,k} t_k \leq v_{e,j} \qquad (48)$$

Lemma 4 *Problem 4 has a solution if and only if Problem 3 has a solution.*

Proof (Sketch) If Problem 4 has a solution, then that is also a solution of Problem 3. If Problem 3 has a solution $\tilde{q}_{i,k}$, $\tilde{t}_k$, then we proceed as follows to construct a solution for the reduced problem. For each k where $\tilde{q}_{j,k}$ is not as prescribed by (45) by constraint (41) we must have

$$\tilde{q}_{j,k} < \min\left(\bar{q}_j, \bar{q}_{tgt} - \sum_{u \in G_k \setminus \{j\}} \underline{q}_u\right)$$

but then we can take

$$\tilde{q}'_{j,k} = \min\left(\bar{q}_j, \bar{q}_{tgt} - \sum_{u \in G_k \setminus \{j\}} \underline{q}_u\right)$$

$\tilde{t}'_k = \tilde{t}_k \tilde{q}_{j,k} / \tilde{q}'_{j,k}$, and then adjust some of the other $\tilde{q}_{i,k}$ downwards to respect (41). These changes respect all constraints. When applied for all k, this results in a solution for Problem 4. □

6.2 Infeasible Problems Due to Pumping Station Grouping

If Problem 4 has a solution and $\mathcal{G}_j$ is such that for all $G \in \mathcal{G}_j$ the number of elements in G is two, so $|G| = 2$, then Algorithm 1 can be applied. In the Garmerwolde case this situation occurred for several Q_{tgt} ranges. The question whether or not it will find a solution is addressed in Lemma 5.

Algorithm 1 Empty a specific reservoir

Initialisation: Define an indexing function $I : \{1, 2, \ldots, K\}$ on $\mathcal{G}_j$ and take $v_i(0) = v_i$ for $i = 1, 2, \ldots, n$.

$k = 1$

while $v_j(k) > v_{e,j}$ and $k \le K$ do

$\quad q_{i,k} = 0$ for $i \notin G$

$\quad$ Let i_k the unique number such that $G(k) = \{i_k, j\}$

$\quad q_{j,k} = \min\left(q_{\max,j}, \max Q_{\text{tgt}} - q_{\min,i_k}\right)$

$\quad q_{i_k,k} = \max\left(q_{\min,i_k}, \min Q_{\text{tgt}} - q_{j,k}\right)$

$\quad t_k = \min\left(\frac{v_{i_k}(k) - \underline{v}_{i_k}}{q_{i_k,k}}, \frac{v_j(k) - \underline{v}_j}{q_{j,k}}\right)$

$\quad$ for $i \in G$ do

$\quad\quad v_i(k+1) = v_i(k) - q_{i,k} t_k$

If $v_j(k) > v_{e,j}$ report failure

Lemma 5 *If Problem 4 has a solution and $\mathcal{G}_j$ is such that for all $G \in \mathcal{G}_j$ the number $|G|$ of elements in G is 2, then Algorithm 1 will find a solution to Problem 4.*

Proof Suppose there is a solution $\tilde{q}_{i,k}$, $\tilde{t}_k$, but the above algorithm does not find a solution. Suppose k_0 is the first index where the solution diverges from the values given in the algorithm. If

$$\tilde{q}_{j,k_0} < q_{j,k}$$

then a new solution can be formulated by taking $\tilde{q}'_{j,k_0} = q_{j,k}$, $\tilde{q}'_{i_{k_0},k_0} = q_{i_k,k}$, $\tilde{t}'_k = t_k$ and reducing the run times $\tilde{t}_{k+1}, \tilde{t}_{k+2}, \ldots, \tilde{t}_K$ to compensate for the additional discharge from j. This can then be repeated for all $k > k_0$ to show that the algorithm does in fact produce a solution. □

Note that a stricter and harder to verify condition is obtained if all t_j must be multiples of a given time step.

7 Conclusion

The problem of evening out the variations in flow to a WWTP while avoiding sedimentation turned out more complex than expected. This was due to the ease with which an infeasible configuration can be created. Simplified problems were derived in the hope that their feasibility would be easier to check. For some of the problems equivalence could be shown. Based in one of the simplified problems a test was developed that will diagnose the problem for a subset of the possible setups that was particularly relevant to the Garmerwolde case. Restrictions in minimum pump run times are likely to create even more problems, so further study on how to make sure the problem statement is feasible is necessary.

References

1. Dyckhoff, H.: A typology of cutting and packing problems. Eur. J. Oper. Res. **44**(2):145–159 (1990). https://doi.org/10.1016/0377-2217(90)90350-k. ISSN 0377-2217
2. NLingenieurs Sewer Systems Workgroup: Sewer Systems Module for Higher Professional Education. KIVI-NIRIA, The Hague, The Netherlands (2009)
3. Perkins, J.R., Kumark, P.R.: Stable, distributed, real-time scheduling of flexible manufacturing/assembly/disassembly systems. IEEE Trans. Autom. Control. **34**(2):139–148 (1989). https://doi.org/10.1109/9.21085. ISSN 0018-9286
4. van Nooijen, R., Kolechkina, A.: A controlled sewer system should be treated as a sampled data system with events. IFAC-PapersOnLine **51**(16):61–66 (2018). https://doi.org/10.1016/j.ifacol.2018.08.011. ISSN 2405-8963. http://www.sciencedirect.com/science/article/pii/S2405896318311248. 6th IFAC Conference on Analysis and Design of Hybrid Systems ADHS 2018
5. van Nooijen, R.R., Kolechkina, A.: Speed of discrete optimization solvers for real time sewer control. Urban Water J. **10**(5):354–363 (2013). https://doi.org/10.1080/1573062x.2013.820330. ISSN 1744-9006
6. van Nooijen, R.R., Kolechkina, A.G.: Realizing steady supply to a treatment plant from multiple sources. IFAC-PapersOnLine **49**(9):29–32 (2016). https://doi.org/10.1016/j.ifacol.2016.07.483. 6th IFAC Symposium on System Structure and Control, SSSC, 2016

Why Burgers Equation: Symmetry-Based Approach

Leobardo Valera, Martine Ceberio and Vladik Kreinovich

Abstract In many application areas ranging from shock waves to acoustics, we encounter the same partial differential equation known as the Burgers' equation. The fact that the same equation appears in different application domains, with different physics, makes us conjecture that it can be derived from the fundamental principles. Indeed, in this paper, we show that this equation can be uniquely determined by the corresponding symmetries.

1 Formulation of the Problem

Burgers' equation is ubiquitous. In many application areas ranging from fluid dynamics to nonlinear acoustics, gas dynamics, and dynamics of traffic flows, we encounter the Burgers' equation; see, e.g., [4, 5]:

$$\frac{\partial u}{\partial t} + u \cdot \frac{\partial u}{\partial x} = d \cdot \frac{\partial^2 u}{\partial x^2}. \tag{1}$$

In particular, our interest in this equation comes from the use of these equations for describing shock waves; see, e.g., [2, 3].

Is there a common explanation for this empirical ubiquity? The fact that the Burgers' equation naturally appears in many different areas seems to indicate that this equation reflects some fundamental ideas, and not just ideas related to liquid or gas dynamics.

L. Valera · M. Ceberio · V. Kreinovich (✉)
University of Texas at El Paso, El Paso, TX 79968, USA
e-mail: vladik@utep.edu

L. Valera
e-mail: leobardovalera@gmail.com

M. Ceberio
e-mail: mceberio@utep.edu

© Springer Nature Switzerland AG 2020
M. Ceberio and V. Kreinovich (eds.), *Decision Making under Constraints*,
Studies in Systems, Decision and Control 276,
https://doi.org/10.1007/978-3-030-40814-5_26

What we do in this paper. In this paper, we show that indeed, the Burgers' equation can be determined from fundamental principles.

2   Let Us Use Symmetries

Why symmetries. How do we make predictions in general? We observe that, in several situations, a body left in the air fell down. We thus conclude that in similar situations, a body will also fall down. Behind this conclusion is the fact that there is some similarity between the new and the old situations. In other words, there are transformations that transform the old situation into a new one—under which the physics will be mostly preserved, i.e., which form what physicists call *symmetries*.

In the falling down example, we can move to a new location, we can rotate around—the falling process will remain. Thus, shifts and rotations are symmetries of the falling-down phenomena.

In more complex situations, the behavior of a system may change with shift or with rotation, but the equations describing such behavior remain the same.

So, let us use symmetries. Symmetries are the fundamental reason why we are capable of predictions. Not surprisingly, symmetries have become one of the main tools of modern physics; see, e.g., [1]. Let us therefore use symmetries to explain the ubiquity of the Burgers' equation.

Which symmetries should we use. Numerical values of physical quantities depend on the measuring unit. For example, when we measure distance x first in meters and then in centimeters, the quantity remains the same, but it numerical values change: instead of the original value x, we get $x' = \lambda \cdot x$ for $\lambda = 100$.

In many cases, there is no physically selected unit of length. In such cases, it is reasonable to require that the corresponding physical equations be invariant with respect to such change of measuring unit, i.e., with respect to the transformation $x \to x' = \lambda \cdot x$.

Of course, once we change the unit for measuring x, we may need to change related units. For example, if we change a unit of current I in Ohm's formula $V = I \cdot R$, for the equation to remain valid we need to also appropriately change, e.g., the unit in which we measure voltage V.

In our case, there seems to be no preferred measuring unit, so it is reasonable to require that the corresponding equation be invariant under transformations $x \to \lambda \cdot x$ if we appropriately change measuring units for all other quantities.

3   What Are the Symmetries of the Burgers' Equation

We want to check if, for every λ, once we combine the re-scaling $x \to x' = \lambda \cdot x$ with the appropriate re-scalings $t \to t' = a(\lambda) \cdot t$ and $u \to u' = b(\lambda) \cdot u$, for some $a(\lambda)$ and $b(\lambda)$, the Burgers' equation (1) will preserve its form.

By keeping only the time derivative in the left-hand side of the equation, we get an equivalent form of the Burgers' equation in which this time derivative is described as a function on the current values of u:

$$\frac{\partial u}{\partial t} = -u \cdot \frac{\partial u}{\partial x} + d \cdot \frac{\partial^2 u}{\partial x^2}. \tag{2}$$

After the transformation, e.g., the partial derivative $\dfrac{\partial u}{\partial t}$ is multiplied by $\dfrac{b(\lambda)}{a(\lambda)}$:

$$\frac{\partial u'}{\partial t'} = \frac{b(\lambda)}{a(\lambda)} \cdot \frac{\partial u}{\partial t},$$

and, more generally, the equation (2) gets transformed into the following form:

$$\frac{b(\lambda)}{a(\lambda)} \cdot \frac{\partial u}{\partial t} = -b(\lambda) \cdot \frac{b(\lambda)}{\lambda} \cdot u \cdot \frac{\partial u}{\partial x} + d \cdot \frac{b(\lambda)}{\lambda^2} \cdot \frac{\partial^2 u}{\partial x^2}. \tag{3}$$

Dividing both sides of this equation by the coefficient $\dfrac{b(\lambda)}{a(\lambda)}$ at the time derivative, we conclude that

$$\frac{\partial u}{\partial t} = -\frac{b(\lambda) \cdot a(\lambda)}{\lambda} \cdot u \cdot \frac{\partial u}{\partial x} + d \cdot \frac{a(\lambda)}{\lambda^2} \cdot \frac{\partial^2 u}{\partial x^2}. \tag{4}$$

By comparing the Eqs. (2) and (4), we conclude that they are equivalent if the coefficients at the two terms in the right-hand side are the same, i.e., if $\dfrac{b(\lambda) \cdot a(\lambda)}{\lambda} = 1$ and $\dfrac{a(\lambda)}{\lambda^2} = 1$. The second equality implies that $a(\lambda) = \lambda^2$, and the first one, that $b(\lambda) = \dfrac{\lambda}{a(\lambda)} = \lambda^{-1}$.

Thus, the Burgers' equation is invariant under the transformation $x \to \lambda \cdot x$, $t \to \lambda^2 \cdot t$, and $u \to \lambda^{-1} \cdot u$.

4 Burgers' Equation Can Be Uniquely Determined by Its Symmetries

Formulation of the problem. Let us consider a general equation in which the time derivative of u depends on the current values of u:

$$\frac{\partial u}{\partial t} = f\left(u, \frac{\partial u}{\partial x}, \frac{\partial^2 u}{\partial x^2}, \dots\right). \tag{5}$$

Here, we assume that the function f is analytical, i.e., that it can be expanded into Taylor series

$$f\left(u, \frac{\partial u}{\partial x}, \frac{\partial^2 u}{\partial x^2}, \dots\right) =$$

$$\sum_{i_0, i_1, \dots, i_k} a_{i_1 \dots i_k} \cdot u^{i_0} \cdot \left(\frac{\partial u}{\partial x}\right)^{i_1} \cdot \left(\frac{\partial^2 u}{\partial x^2}\right)^{i_2} \cdot \dots \cdot \left(\frac{\partial^k u}{\partial x^k}\right)^{i_k}, \tag{6}$$

where $i_0, i_1, \dots, i_k$ are non-negative integers.

We are looking for all possible cases in which this equation is invariant under the transformation $x \to \lambda \cdot x$, $t \to \lambda^2 \cdot t$, and $u \to \lambda^{-1} \cdot u$.

Analysis of the problem. Under the above transformation, the left-hand side of the Eq. (5) is multiplied by $\dfrac{\lambda^{-1}}{\lambda^2} = \lambda^{-3}$. On the other hand, each term in the expansion (6) of the right-hand side of the formula (5) is multiplied by

$$\left(\lambda^{-1}\right)^{i_0} \cdot \left(\lambda^{-2}\right)^{i_1} \cdot \left(\lambda^{-3}\right)^{i_2} \cdot \dots \cdot \left(\lambda^{-(k+1)}\right)^{i_k}, \tag{7}$$

i.e., by λ^{-D}, where we denoted

$$D = i_0 + 2 \cdot i_1 + 3 \cdot i_2 + \dots + (k+1) \cdot i_k. \tag{8}$$

The equation is invariant if the left-hand side and right-hand side are multiplied by the same coefficient, i.e., if $D = 3$. Thus, in the invariant case, we can have only terms for which

$$i_0 + 2 \cdot i_1 + 3 \cdot i_2 + \dots + (k+1) \cdot i_k = 3. \tag{9}$$

Here, the values $i_0, \dots, i_k$ are non-negative integers. So, if we had $i_j > 0$ for some $j \geq 3$, i.e., $i_j \geq 1$, the left-hand side of the formula (9) would be greater than or equal to $j + 1 \geq 4$, so it cannot be equal to 3. Thus, in the invariant case, we can only have values i_0, i_1, and i_2 possibly different from 0. In this case, the formula (9) takes a simplified form

$$i_0 + 2 \cdot i_1 + 3 \cdot i_2 = 3. \tag{10}$$

If $i_2 > 0$, then already for $i_2 = 1$, the left-hand side of (1) is greater than or equal to 3, so in this case, we must have $i_2 = 1$ and $i_0 = i_1 = 0$. This leads to the term $d \cdot \dfrac{\partial^2 u}{\partial x^2}$ for some d.

Let us consider the remaining case $i_2 = 0$. In this case, the Eq. (10) has the form $i_0 + 2 \cdot i_1 = 3$. Since $i_0 \geq 0$, we have $2i_1 \leq 3$, so we have two options: $i_1 = 0$ and $i_1 = 1$.

- For $i_1 = 0$, we have $i_0 = 3$, so we get a term proportional to u^3.
- For $i_1 = 1$, we get $i_0 = 1$, so we get a term proportional to $u \cdot \dfrac{\partial u}{\partial x}$.

Thus, we arrive at the following conclusion.

Conclusion: which equations have the desired symmetry. We have shown that any equation invariant under the desired symmetry has the form

$$\frac{\partial u}{\partial t} = a \cdot u \cdot \frac{\partial u}{\partial x} + d \cdot \frac{\partial^2 u}{\partial x^2} + b \cdot u^3. \tag{11}$$

By changing the unit of x to $|a|$ times smaller one (and maybe changing the direction of x), we can make the coefficient a to be equal to -1:

$$\frac{\partial u}{\partial t} = -u \cdot \frac{\partial u}{\partial x} + d \cdot \frac{\partial^2 u}{\partial x^2} + b \cdot u^3. \tag{12}$$

This is *almost* the Burgers' equation, the only difference is the new term $b \cdot u^3$.

This term can be excluded is we take an additional assumption that if the situation is spatially homogeneous, i.e., if $\dfrac{\partial u}{\partial x} \equiv 0$, then there is no change in time, i.e., $\dfrac{\partial u}{\partial t} = 0$.

How can we justify this additional requirement? Suppose that this requirement is not satisfied; then, in the homogeneous case, we have

$$\frac{du}{dt} = b \cdot u^3,$$

i.e., equivalently,

$$\frac{du}{u^3} = b \cdot dt$$

and, after integration, $u^{-2} = A \cdot t + B$ for some A and B. Thus, we have

$$u = \frac{1}{\sqrt{A \cdot t + B}}.$$

If we want to avoid such a spontaneous increase or decrease, then, from the invariance requirement, we get only the Burgers' equation.

Acknowledgements This work was supported in part by the US National Science Foundation grant HRD-1242122.

References

1. Feynman, R., Leighton, R., Sands, M.: The Feynman Lectures on Physics. Addison Wesley, Boston, MA (2005)
2. Valera, L.: Contributions to the solution of large nonlinear systems via model-order reduction and interval constraint solving techniques. Master's Thesis, Computational Science Program, The University of Texas at El Paso (2015)
3. Valera, L., Ceberio, M.: Model-order reduction using interval constraint solving techniques. J. Uncertain Syst. **11**(2), 84–103 (2017)
4. Zwillinger, D. (ed.): CRC Standard Mathematical Tables and Formulae. CRC Press, Boca Raton (1995)
5. Zwillinger, D.: Handbook of Differential Equations. Academic Press, Boston (1997)

Working on One Part at a Time Is the Best Strategy for Software Production: A Proof

Francisco Zapata, Maliheh Zargaran and Vladik Kreinovich

Abstract When a company works on a large software project, it can often start recouping its investments by selling intermediate products with partial functionality. With this possibility in mind, it is important to schedule work on different software parts so as to maximize the profit. These exist several algorithms for solving the corresponding optimization problem, and in all the resulting plans, at each moment of time, we work on one part of software at a time. In this paper, we prove that this one-part-at-a-time property holds for all optimal plans.

1 Formulation of the Problem

It is possible to start earning money before the whole software package is released. When a company designs a software package, usually, it does not have to wait until the whole package is fully functional to profit from sales: the company can often start earning money once some useful features are implemented.

As an example, let us consider a company that designs a package for all kinds of numerical computations, including solving systems of equations, optimization, etc. Instead of waiting until all the parts of the software are ready, the company can first release—and start selling—the parts that solve systems of equations. Thus, the company can start earning money before the whole package is ready for use.

This possibility is critical. Indeed, for large software packages, full design can take years. So, the possibility to recoup at least some of the original investment earlier makes such long-term projects more acceptable to managers and shareholders—and thus, makes these projects more probable to be approved; see, e.g., [1].

F. Zapata · M. Zargaran · V. Kreinovich (✉)
University of Texas at El Paso, El Paso, TX 79968, USA
e-mail: vladik@utep.edu

F. Zapata
e-mail: fazg74@gmail.com

M. Zargaran
e-mail: mzargaran@miners.utep.edu

© Springer Nature Switzerland AG 2020
M. Ceberio and V. Kreinovich (eds.), *Decision Making under Constraints*,
Studies in Systems, Decision and Control 276,
https://doi.org/10.1007/978-3-030-40814-5_27

In view of this possibility, what is the optimal release schedule for different parts? How can we best take into account the possibility to earn money before the package is fully ready, when only some parts of it are ready? What is the optimal schedule for releasing different parts? In what order should we work on them?

Let us formulate this problem in precise terms. The software projects consist of several parts. Some of these parts depend on others, in the sense that we cannot design one part until the other part is ready.

For example, many numerical techniques for solving systems of *nonlinear* equations use linearization, and thus, solve systems of *linear* equations at different stages. So:

- in order to design a part for solving systems of nonlinear equations,
- we need to have available a part for solving systems of linear equations.

This dependency relation makes the set of all parts into a partially ordered set, i.e., in other words, into a directed acyclic graph.

For each part i, we know the overall effort e_i (e.g., in man-hours) that is needed to design this part (by utilizing, if needed, all the parts on which it depends). We also know the profit p_i that we can start earning once this part is released – by adding this part's functionality to whatever we were selling before.

So, if we release part i at time t_i, then by some future moment of time T, selling this part will bring us the profit of $p_i \cdot (T - t_i)$.

The question is: how can we organize our work on different parts so as to maximize the resulting overall profit.

What is known. In [1], several semi-heuristic strategies are described that lead to optimal (or at least close-to-optimal) release schedules.

Interestingly:

- while it is, in principle, possible for the company to work on several parts at a time,
- in all known optimal schedules, the design is performed one part at a time.

What we do in this paper. In this paper, we show that the above empirical fact is not a coincidence: we will actually prove that in the optimal schedule, we *should* always work on one part at a time.

2   Main Result

Exact formulation of the main result that we prove in this paper. What we will prove is that *for each planning problem, there is an optimal schedule in which we always work on one part at a time*.

Discussion. This result does not necessarily means that in *all* optimal scheduled, we work on one part at a time. Let us give a simple example.

Suppose that the software consists of three parts:

- The first two parts are independent and require the same time $t_1 = t_2$.
- The third part depend on the first two parts.

Suppose also that there is no market need for Part 1 or for Part 2, only for the final Part 3. In other words, we assume that $p_1 = p_2 = 0$.

In this case, it is reasonable to conclude that the following schedule is optimal:

- first, we work on Part 1; this take time t_1;
- then, we concentrate all or efforts on Part 2; this also takes time $t_1 = t_2$;
- after this, we work on Part 3; this takes time t_3.

So, by time $2t_1 + t_3$, we get the product that we can start selling.

Alternatively, we can use a different schedule:

- first, we split the team into two equal sub-teams, with one sub-team working on Part 1, and the other sub-team working on Part 2; since designing each part takes time $t_1 = t_2$ for the whole team, it will take twice longer for the twice-smaller sub-teams;
- after the time $2t_1$, both Part 1 and Part 2 are ready, so we can start working on Part 3.

At the end, after time $2t_1 + t_3$, we will get the ready-to-sell product.

Comment. In this example, at least one of the parts does not bring any profit, i.e., has $p_i = 0$. We will see that if each part can bring some profit, i.e., if $p_i > 0$ for all i, then such examples are not possible, and in *all* optimal schedules, we work on one part at a time.

3 Proof

Transformation and its properties. Let us start with an optimal schedule in which at some moment of time, we work on several parts at the same time.

Let t_f be the first moment of time with this property. If any part which is ready by this moment is not yet released in the original optimal plan, we can release it right away and thus provide the additional income from selling this part. Thus, without losing generality, we can assume that:

- all the parts which are released after moment t_f
- are not yet fully ready for release at the moment t_f.

Let t_r be the first moment of time after t_f at which one of the parts is being released, and let j be the number of the part that is released at moment t_r.

Then, we can modify the original schedule as follows.

- Right after the moment t_f, we concentrate all our efforts on this Part j—and all the efforts aimed at other parts are performed after that.
- By the time t_r, we spend the exact amount of efforts on all the parts as before.

- However, since in the original plan, around moment t_f, we were also working on some other parts, this concentration means that we can now release Part j earlier—while preserving the release dates for all other parts.

Thus, in comparison with the original plan, we can earn more (or at least same amount of) profit, since we started selling Part j earlier.

Two possibilities. The profit p_j is always non-negative. So, we have two options: $p_j > 0$ and $p_j = 0$. Let us consider these two options one by one.

Case when $p_j > 0$. If $p_j > 0$, we indeed get more profit—which contradicts to our assumption that the original plan was optimal. Thus, when $p_i > 0$ for all parts, in the optimal plan, we cannot have moments of time at which we work on several parts at a time—which is exactly what we are trying to prove.

Case when $p_j = 0$. If $p_j = 0$, then we do not get any contradiction, but we get a new optimal plan in which:

- either we always work on one pat at a time
- or the first moment of time at which we work on several parts at a time moves further, to some moment $t'_f > t_f$—since in the vicinity of the moment t_f, we now work on only one part.

In the new plan, we have at least one fewer part on which we work simultaneously with others. Please note that in the new plan, when we work on Part j, we always work *only* on this part. Indeed:

- This has been true—and remains true—before the moment t_f, since t_f is the first moment at which we work on several parts at a time.
- We have also designed a new plan in such a way that this is also true for all moments $t \geq t_f$.

If the resulting plan is not the desired one, we can apply similar transformations. If in the new plan, we still have moments of time when we simultaneously work on several parts, then, by applying the same transformation to the new first-such-moment t'_f, we get yet another optimal plan for which:

- either there is no time when we several parts at a time,
- or the first moment t''_f when this happens is even further away.

Let us prove that this process converges to the desired plan. At each such step, we delete at least one part from the list of parts on which we work simultaneously with others. Thus, after at most as many steps as there are parts, we will get an optimal plans in which:

- no such parts remain and thus
- we always work on one part at a time.

This final optimal plan is exactly the plan whose existence we wanted to prove. Thus, our main result is proven.

Acknowledgements This work was supported in part by the US National Science Foundation grant HRD-1242122.

Reference

1. Denne, M., Cleland-Huang, J.: Software by Numbers: Low-Risk, High-Return Development. Prentice Hall, Upper Saddle River, New Jersey (2004)

Printed in the United States
by Baker & Taylor Publisher Services